C000193934

TALES FROM BARRA

TOLD BY THE CODDY

Tales From
BARRA
TOLD BY THE CODDY

JOHN MACPHERSON,
NORTHBAY, BARRA, 1876–1955

Foreword by
Compton Mackenzie

––––––––

Introduction and Notes
by John Lorne Campbell

ORIGIN

This edition first published in 2018 by
Birlinn Origin, an imprint of
Birlinn Limited
West Newington House
10 Newington Road
Edinburgh
EH9 1QS

www.birlinn.co.uk

First published by Birlinn Ltd in 1992
Reprinted 2001, 2004, 2008, 2014

Copyright © the Estate of John MacPherson 1992

All rights reserved. No part of this publication may be reproduced, stored
or transmitted in any form, or by any means, electronic, mechanical or pho-
tocopying, recording or otherwise, without the express written permission
of the publisher.

ISBN 978 1 91247 617 6
eBook ISBN 978 0 85790 977 0

British Library Cataloguing-in-Publication Data
A catalogue record for this book is available from the British Library

Typeset by Geethik Technologies, India
Printed and bound by Clays Ltd, St Ives plc

Contents

*Wire recordings of these stories made from the Coddy by the editor are in existence.

Foreword

I met the Coddy first at the Inverness Mòd in 1928 and made up my mind immediately that I must lose no time in visiting Barra in order to enjoy more of as good company as I have ever known. This collection of his tales with which my old friend John Campbell has put us in his debt needs no bush of words from me to proclaim the quality of such a vintage: it is evident. There are happily still many who will read these tales and hear the voice of the Coddy telling them, but I am sadly aware of my own inability to evoke him on the page for those who knew him not. His wit was constant, and, though usually inspired by the humour of the moment, he was able to retain it in his correspondence. The briefest letter from the Coddy had always a phrase to make it memorable, and I never received a letter from him but I wished I were with him, and that is a precious rarity in my correspondence today.

The Coddy had an infallible sense of a man's worth. I never knew him 'put his money' on an impostor. The socially pretentious, the bore, the sponger, the sentimentalist, the romantic liar never misled him into accepting their own opinion of themselves: he had a deadly objectivity. He possessed a remarkable gift which he shared with the Cheshire Cat of being able to disengage himself from present company while apparently he was still there. In his case what remained was not a smile but a pair of intensely blue eyes of which those who were familiar with them knew that the owner was no longer there.

Self-possession was one of the Coddy's characteristics, but it had not been granted to him by a good fairy at birth. We were driving round Barra once, and at Allt the Coddy stopped the car.

'I want to show you something,' he said to me when we alighted. 'This was the very spot where I made up my mind when I was young that I was as good as other people. I had always felt awkward when I was selling the fish and I said to myself, 'You can sell fish against anybody, and you must understand that and not feel awkward.

And I went on my way to Castlebay and from that moment I felt I was as good as anybody.'

Yet there were moments when that self-consciousness of youth he conquered once upon a time would assert itself. I remember a Corpus Christi procession in which he and I were taking part. The Coddy was to carry the crucifix at the head of the procession, and as he came out of the sacristy in cassock and cotta he noticed that Ninian, his youngest son, was laughing with another altar-boy.

The Coddy was not prepared for me to see him an object of mirth for the youth of Northbay. He must laugh at himself first.

'Direct from the Vatican,' he said to me, in that voice of mock solemnity which those who ever heard it will so much regret that they will never be able to hear again.

One of the innumerable pleasures of the years I lived in Barra was visiting with the Coddy the small isles about it that were no longer inhabited, because he was able to conjure up from the past those who within his own memory had been the life of them, Fuday, Fiaray, Hellisay, Sandray, Pabbay, Mingulay, Berneray ... every name like the gnomon of a sun-dial records for me only the sunny hours, and on every one of them it is always a blue summer sky with the *machair* in fullest flower and reminiscence from the Coddy flowing like the tide.

One curious vice the Coddy had (I do not apologise for the strength of the word), and that was to suppose that whisky was improved by diluting it with fizzy lemonade. I recall a visit to Barra by Lochiel, Charles Tinker and Alasdair Fraser of Lovat as repre sentatives of the Inverness-shire County Council of which the Coddy was the Member for the north end of the island. Hospitable always but on this occasion anxious to be particularly hospitable, the Coddy pressed upon them the suitability of the moment for refreshment. None was loth, and the Coddy retired to prepare the drams which he was determined should be drams de luxe. I can see now the expression of horror and incredulity upon the faces of his three prized guests as they tasted the liquor.

'My God, what's this?' Charlie Tinker gasped.

And the Coddy, raising both hands in a *sursum corda* of devout benison, said proudly: 'Whisky and lemonade.'

One more story. On a glorious June morning the Coddy and I had gone down to greet Eric Linklater's arrival with his bride for their honeymoon in Barra, and on the way down to the pier I was beckoned to by a citizen of Castlebay with whom the Coddy was not on the most cordial terms. I went into his shop, and later joined the Coddy on the pier.

'Well, Coddy,' I told him, 'I've just been given something you were never given by X–Y–.'

'What was that?'

'Two drams, It's his birthday.'

'A *Dhia!* the Coddy exclaimed in sombre marvel. 'Fancy a man like that to be born in the month of June!'

I shall not hear again the gurgle of the Coddy's pipe nor see the ritual of expectoration that concluded the lighting of it, but as I write these words he is sitting on the other side of the fire as vivid and as much loved a figure in memory as he was in my life. He rests now by the Oitir Mhòr which he loved so dearly, and it is my hope that one day I shall rest near him and other old friends in the last *cèilidh* of all.

COMPTON MACKENZIE

MACNEIL OF BARRA PEDIGREES

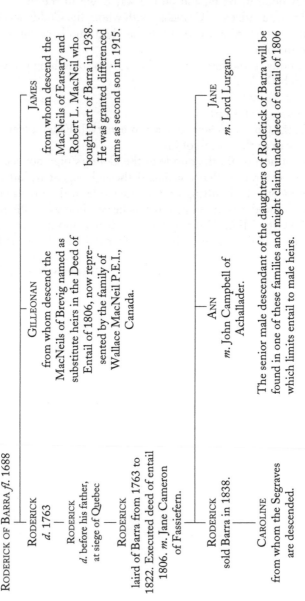

RODERICK OF BARRA *fl.* 1688

RODERICK
d. 1763

RODERICK
d. before his father,
at siege of Quebec

RODERICK
laird of Barra from 1763 to
1822. Executed deed of entail
1806. *m.* Jane Cameron
of Fassiefern.

RODERICK
sold Barra in 1838.

CAROLINE
from whom the Segraves
are descended.

GILLEONAN
from whom descend the
MacNeils of Brevig named as
substitute heirs in the Deed of
Entail of 1806, now repre-
sented by the family of
Wallace MacNeil P.E.I.,
Canada.

ANN
m. John Campbell of
Achallader.

JAMES
from whom descend the
MacNeils of Earsary and
Robert L. MacNeil who
bought part of Barra in 1938.
He was granted differenced
arms as second son in 1915.

JANE
m. Lord Lurgan.

The senior male descendant of the daughters of Roderick of Barra will be
found in one of these families and might claim under deed of entail of 1806
which limits entail to male heirs.

N.B. – There is a Barra tradition that Gilleonan was older than Roderick but was passed over. In a similar way it appears
that the MacNeils of Vatersay are descended from Niall Uibhisteach who lost the succession to a usurping illegitimate
brother around 1615.

CODDY'S FAMILY TREE

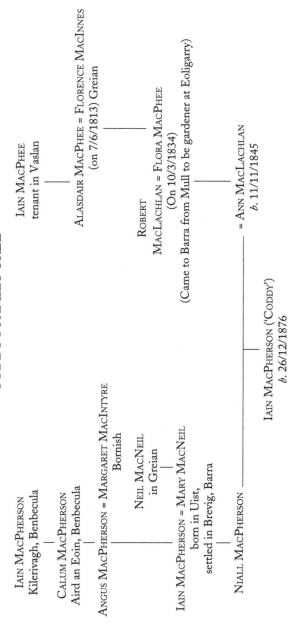

IAIN MACPHEE
tenant in Vaslan

ALASDAIR MACPHEE = FLORENCE MACINNES
(on 7/6/1813) Greian

ROBERT
MACLACHLAN = FLORA MACPHEE
(On 10/3/1834)
(Came to Barra from Mull to be gardener at Eoligarry)

= ANN MACLACHLAN
b. 11/11/1845

IAIN MACPHERSON ('CODDY')
b. 26/12/1876

IAIN MACPHERSON
Kilerivagh, Benbecula

CALUM MACPHERSON
Aird an Eoin, Benbecula

ANGUS MACPHERSON = MARGARET MACINTYRE
Bornish

NEIL MACNEIL
in Greian

IAIN MACPHERSON = MARY MACNEIL
born in Uist,
settled in Brevig, Barra

NIALL MACPHERSON

Angus MacPherson had brothers;
Cathelus, grandfather of Roderick MacMillan, Gerinish.
Donald, who lived at Griminish, and had a daughter Catherine, mother of Lachy Bàn, the famous piper.

Introduction

My first meeting with the Coddy – a nickname bestowed in boy-hood and persisting, like so many Barra nicknames, throughout a lifetime – was a brief one in August 1928. Later the same year I had the pleasure of travelling with him by bus on the old Loch Ness-side road from Fort William to Inverness, where we were both go-ing to attend a Mòd, and Coddy, who represented the northern half of Barra on the Inverness-shire County Council, was also to attend a Council meeting. An invitation to return to Barra for the purpose of studying colloquial Gaelic was warmly extended. My only regret now is that I was unable to take advantage of it until 1933.

Coddy was in appearance rather short, thick-set and Napoleon-ic; he had an extremely fine-looking head and was quick of move-ment – and of speech, whether in English or Gaelic. His MacPher-son forebears came originally from Benbecula. Sixty years ago the late Fr Allan McDonald of Eriskay recorded a South Uist saying, '*Geurainich Chlann Mhuirich*', '*Geuraineach*, a smart-tongued fellow. The MacPhersons or Curries or MacVurichs are considered sharp on their tongues and apt scholars.' This was very true of the Coddy, one of whose chief characteristics was aptness of speech, both in Gaelic and English, and a talent for both anecdote and diplomacy which might, as I have heard it said, have made him Mayor of a large American city, had he been a citizen of that republic. As it was, his talents did much for his native island, and delighted a very large circle of friends.

There is considerable interest in Coddy's family tree. It belies completely the popular notion that the inhabitants of the Outer Hebrides formed isolated and inbred communities. On the pater-nal side his people came from Benbecula and South Uist, on the maternal from the Island of Mull, both sides eventually marrying into Barra families after settling on the island, so that both the Coddy's grandmothers, Mary MacNeil and Flora MacPhee, were

Barra-born. But perhaps his descent can best be described in the words of his eldest daughter, Miss Kate MacPherson, into which I have interpolated some dates and other information obtained from the Barra Baptismal Register and from Mr Roderick MacMillan, Gerinish, South Uist, one of Coddy's cousins.

Miss Kate MacPherson writes: 'My father's maternal grandfather, Robert MacLachlan, was gardener at Aros Castle in Mull and came from there to be gardener for Colonel MacNeil of Barra at Eoligarry. When telling this, my father always said that his grandfather came to Barra 'after the Napoleonic wars,' but apart from that I regret I cannot give a date. Robert MacLachlan became a Catholic and married Flora MacPhee' [daughter of Alexander Macphee, Vaslain, and Flora MacInnes, Greian, his wife, born 6th September 1815. The marriage took place at Eoligarry on 10th March 1834 and was performed by Fr Neil MacDonald]. 'Robert and Flora subsequently moved to Talisker in Skye, but later returned to Barra and settled on a croft at Buaile nam Bodach. My great-grandmother brought her children from Skye to be baptized at Craigstone.' [Of the five youngest children of Robert MacLachlan and Flora MacPhee, Archibald was born on 3rd April 1840 and baptized on 23rd August of the same year; Janet was born on 4th November 1841 and baptized on 22nd June 1844; and Margaret born on 15th April 1843, Ann (Coddy's mother), born on 11th November 1845, and Neil, born 12th June 1848, were all baptized together on 21st June 1848. It therefore looks as if Robert MacLachlan and his family left Barra not long after the sale of the estate to Colonel Gordon in 1838, when Eoligarry ceased to be a laird's residence, and returned to Barra in 1848, when the potato famine may have made things difficult on Skye.]

'One of their daughters, Ann, married Neil MacPherson – Niall mac Iain 'ic Aonghuis 'ic Caluim 'ic Iain. I am sorry I cannot go further with the *sloinneadh*, but my father's cousin Roderick Mac Pherson tells me that my father could go back fourteen or fifteen generations. Ann and Neil settled with my great-grandfather, Robert MacLachlan, on the croft at Buaile nam Bodach, and there my father was born on 26th December 1876.

'Now as to the MacPhersons. Iain and his father Angus were both joiners and came from South Uist to work on the priest's house at Craigstone, Taigh a' Ghearraidh Mhóir (the house of the big wall).'

[The priest in question was Fr Angus MacDonald, who was in Barra from 1805 to 1825, when he became Rector of the Scots College at Rome. Some of his correspondence is printed in the *Book of Barra*. His account book shows that Angus MacPherson came in April 1819 to start work on his house, bringing his wife Mary MacIntyre and his son Iain with him. Another son, Lachlan MacPherson, was born on Barra and baptized on 3rd April 1820, when his parents were described in the Barra baptismal register as 'natives of South Uist, now residing in this parish,' and the godfather was Allan MacArthur of the sloop *Maid*. Angus MacPherson, besides being a well-known joiner, was a bard. Here are two of his best-known songs, as taken down by Fr Allan McDonald in 1887 and 1897:

'The following comical song was composed by a carpenter of the name of McPherson, commonly called Aonghus mac Caluim 'ic Iain:

Tha Siosalaich is Griogalaich tric 'ga mo bhòdadh,
Iad trom orm uile, 's mi umhail gu leòr dhaibh,
Air son siochaire botuil bhith 'ga chosg' san taigh òsda,
'S mi fhin bhith 'ga chosnadh le locraichean gròbaidh,
                                    Mo chailin donn òg.

Chuir mi fios air an t-sagart bha stad anns an Iochdar,
Bho'n a bha e n'a dhotair gu socair dhomh dhianamh;
'S nuair bha e 'gam shealltainn, bha mo cheann-sa gun riaghailt,
'S ann thuirt e 'Tha'm bàs ort, cha n-fhàg e thu 'n dìochainn,'
                                    Mo chailin donn òg.

'S tric tha mi smaointinn na daoine nach maireann
A dh'òladh 's a phàigheadh 's nach fhàgadh mi falamh
Nam biodh fios agam fhìn nach eil sìth aig an anam
'S e 'n obair nach fhiach a bhith dianamh an drama,
                                    No idir 'ga h-òl.

Tha mo bhean-sa air a marbhadh a' falbh feadh an fhearainn,
A' tional 's ag iarraidh gach sian dhomh gu fallus,

16

A dh'aindeoin a pianadh 's a riasladh 's a caithris,
A dh'aindeoin a saothair, tha 'n saor gu bhith thairis,
Mo chailin donn òg.

Ach ma leighiseas mo shùilean, bheir mi 'n ionnsaigh so fhathast,
Null far a' chaoil far bheil gaolach nam fearaibh,
Far bheil ceannard na cléire nach leubhadh a' ghainne,
'S beag iùnadh do threud gun dhol ceum ann am mearachd,
Mo chailin donn òg.'

Translation

'Chisholms and MacGregors are often putting me under the pledge, they are all hard on me, though I am obedient enough to them, because I consumed a trifling bottle at the inn, which I earned with grooving planes – My young brown lass (chorus).

I sent for the priest who was living at Iochdar, because he was a doctor, to restore me to health; when he was looking at me, my head was out of order; he said, 'Death is coming for you, he will not leave you forgotten' – My young brown lass.

Often I think of the men once alive, who used to drink and pay and would not leave me empty; if I knew that their souls were not in peace, it would be worthless work to be taking a dram, or drinking one at all.

My wife is worn out going through the land, collecting and seeking every herb that would make me perspire; in spite of her pain and her discomfort and her watching, in spite of her work, the joiner [himself] is nearly done for – My young brown lass.

But if my eyes are cured, I will make the attempt yet to go across the sound [Barra Sound] where is the dearest of men; where there is the chief of the clergy who does not expatiate on scarcity; little wonder your flock does not go one step into error – My young brown lass.'

Fr Allan says that, 'The Siosalach and Griogalach were the Rev. John Chisholm of Bornish and the Rev. James MacGregor of Iochdar. The priest of Barra was the Rev. Neil McDonald who died afterwards at Drimnin, Morvern.'

The priest of Barra must, however, have been the Rev. Angus

MacDonald, for whom Angus MacPherson went to work in 1819 at Craigstone.

The other song is:

'Oran a rinn Aonghus Mac Caluim do sgothaidh a bha 'm Barraidh *[deleted,* Cille Brighide *substituted]* ris an canadh iad "An Cuildheann" – Coolin Hills.

> Hug a rì hu gu gù rìreamh
> Air a' bhàta làidir dhìonach,
> Do Not Faro 's do na h-Innsibh
> Bheir i sgriob o thìr a h-eòlais.
>
> Thug iad an Cuildheann a dh'ainm ort,
> Beanntannan cho àrd 's tha 'n Alba;
> Cuiridh canbhas air falbh iad,
> Ged tha siod 'na sheanchas neònach.
>
> Gaoth an iardheas far an fhearainn,
> 'S ise 'g iarraidh tighinn a Bharraidh;
> Chuir i air an t-sliasaid Canaidh,
> 'S *bheat* i 'n cala gu Maol-Dòmhnaich.
>
> Bàta luchdmhor làidir dìonach
> Gum bu slàn an làmh 'ga dianamh;
> *bheat* i na bha 'n taobh-sa Ghrianaig
> Ach a cunntas liad 'san t—seòl dhi.
>
> Gaoth an iardheas as an Lingidh
> Toiseach lionaidh, struth 'na mhire,
> 'S ise mach iarradh gu tilleadh
> Ach a gillean a bhith deònach.'

Translation

'A song which Angus son of Calum made to a boat which was in Barra (*deleted, and* Kilbride *substituted* – Kilbride is in South Uist) which was called *The Coolins.*

Chorus: *Hug a ri hu gu* verily
> On the strong watertight boat
> To Not Faro and to the Indies
> She will journey from her own country.

They called you "The Coolins," mountains as high as any in Scotland; canvas will set them moving, though that is a strange tale.

The wind is in the south-west coming off the land, as she seeks to come to Barra; she left Canna on her beam, and beat into harbour to Muldonich.

A valuable, strong, watertight boat, may the hand that made her prosper; she beat what boats there are on this side of Greenock, only counting the breadth of the sail in her.

The south-west wind is off the sound [between Barra and Vatersay], flood tide starting, current playing; she would not ask to turn back as long as her lads were willing.'

[A version of eight verses, but lacking the fourth verse given here, was printed by Colm O Lochlainn in *Deoch Slàinte nan Gillean*, in which it is stated that the song was made by the Coddy's great-uncle to the boat in which he and the Coddy's great-grandfather came from Mull to Barra; but this can hardly be the case, for it was the Coddy's maternal great-grandfather Robert MacLachlan who came from Mull to Barra, not Angus MacPherson, the bard, though this boat might have brought them.]

'Iain MacPherson (Angus's son) settled in Brevig, had a boat which he built himself, and was drowned on the Oitir. After his death his widow (Mary MacNeil) was evicted from the house in Brevig and went to live in Bruernish, near Northbay.'

[Iain's son Niall married Ann MacLachlan, as has been stated.]

'When my father left school, he started work lobster fishing with his father, and later worked on Donald William MacLeod's boats line fishing and herring fishing.' Donald William MacLeod was married to the Coddy's aunt. 'Probably about this time he joined the Royal Naval Reserve. One summer he had a very severe attack of pneumonia, and after that did not return to the fishing. Instead, he started working for Joseph MacLean, one of the leading Barra merchants, and became his fish salesman. This work took him to Skye, Lochboisdale, and, of course, to Castlebay (which was then a great herring curing centre). In the off-season he worked at Skallary and Northbay. In 1911 he got a croft at Northbay, built a house and set up a merchant's business of his own, and in 1923 he was appointed Postmaster of Northbay.'

19

I continue the account with a quotation from the obituary printed in the *Oban Times*:

'When cars came to the island he started a car and motor-hiring business with considerable success [actually I believe he had the first car ever imported into Barra, a model T Ford]. As the tourists began to flock to the island he decided to build a boarding house, and this was the "Coddy" in his element and at his very best. He was the genial host, the great story-teller and the charming *fear-an-taighe*. Many will remember pleasant hours spent in Taigh a' Choddy listening to his fund of folk-lore told with feeling and sincerity. Most of these were told in Gaelic, but the "Coddy" was equally fluent in English.

'It was not surprising to hear he had taken part in the film during the shooting of *Whisky Galore* on the island of Barra. In fact, with such a personality he could have had Hollywood at his feet. Indeed, it can be said that it is possible that the author would never have written the book if there had been no "Coddy".

'He served for a number of years on the Barra District Council and also on the County Council of Inverness, where he gave unstinted service to his island.

'One of the many great qualities in his life was his devotion and loyalty to his church. He saw the church, St Barr at Northbay, go up, and he took a keen and leading interest in it all of his days. His fondness of children and his kindness and readiness to help when ever he was approached testified to his true Christian character.

'In June 1944 he received a severe blow in the death of his second son, Neil, who was killed in action while serving with the R.A.F. This upset the Coddy very noticeably, and he was never quite the same again. For the past year he had been confined to the house with cardiac trouble, and the end came peacefully on the morning of 27th February, 1955.

'After Requiem Mass in St Barr's church, Northbay, he was laid to rest in the cemetery at Kilbar in Eoligarry, facing the Oitir Mhór of which he was so fond. To his widow and family of three sons and three daughters his very many friends offer their sincere condolences, and their farewell to the Coddy they express in the native tongue by that final prayer of the church:

Fois shìorruidh thoir dha, a Thighearna,
Agus solus nach dìbir dearrsadh air.'1

Coddy's personality and talents as a host brought him before long a large number of visitors, some of them distinguished ones – their names are preserved in his green book – peers, politicians, officials, descendants of Barra emigrants to Canada and the U.S.A., scholars from Scotland, Norway and Gaelic Ireland, archaeologists, ornithologists, sportsmen and holiday-makers simply seeking a change and a rest in the peaceful unhurried atmosphere of pre-second war Barra, all made their way to the Coddy's, attracted by his vigorous personality and the kindness and hospitality of his wife and family.

I did not know that Coddy's house had become so popular when I returned to study colloquial Gaelic there in the summer of 1933. Conversation with English-speaking visitors, though often interesting, did not further these studies very much. But Coddy was assiduous in his teaching. I got no English from him. 'Abair siod fhathast, Iain,'2 he used to say to me whenever I made an error in pronunciation or used the wrong word. As autumn drew on, and the visitors departed southwards with the puffins and other migrants, we got more time to ourselves to study the language which we both loved.

Twenty-odd years ago Barra was an island where, one felt, time had been standing still for generations. It is always extraordinarily difficult to convey the feeling and atmosphere of a community where oral tradition and the religious sense are still very much alive to people who have only known the atmosphere of the modern ephemeral, rapidly changing world of industrial civilisation. On the one hand there is a community of independent personalities where memories of men and events are often amazingly long (in the Gaelic-speaking Outer Hebrides they go back to Viking times a thousand years ago), and where there is an ever-present sense of the reality and existence of the other world of spiritual and psychic experience; on the other there is a standardised world where people live in a mental jumble of newspaper headlines and B.B.C. news bulletins, forgetting yesterday's as they read or hear today's, worrying themselves constantly about far-away events which they cannot

---

[1] Give him everlasting rest, O Lord, and light that never fails. [2] 'Say that again, John'

possibly control, where memories are so short that men often do not know the names of their grandparents, and where the only real world seems to be the everyday material one. If it be the case that 'Gaelic alone is not enough to keep a man alive' and that therefore the Hebridean world of oral tradition must yield to the encroachment of mass semi-sophistication and anglicised education, so that the islanders be not cheated in the labour markets of the south, that does not mean that victory is going to the 'better' of the two contestants in the struggle, but simply that the material progress of the Islands is being achieved at the cost of cultural impoverishment, which makes one envy the more the Icelanders and Faeroese who have contrived to make the best of both worlds, and are retaining their ancient languages as instruments of modern culture and education, while bringing their material way of life up to date.

In the Barra of Coddy's time, psychic experiences were still sufficiently frequent to maintain the particular character these have always given to the Hebrides, as can be seen from his stories. Material events, such as the evictions, the potato famine, the departure of the last of the old race of lairds in the direct line, the Napoleonic wars and the oppression of the pressgang, none of which happened later than 1851, seemed to be matters of yesterday. The Jacobite rising of 1745 and 1715 felt only a little farther back, and the events of the seventeenth century, the wars between Royalists and Covenanters, and the visits to Barra of the Irish Franciscans (1624–40) and the Vincentians (1652–57), of whom Fr Dermid Dugan was particularly well remembered, seemed only a very little earlier. Behind all this lay memories of the exploits of the old MacNeils of Barra, of the Lords of the Isles, and of the Viking invaders of Scotland and Ireland, who started coming in the ninth century: the island of Fuday in Barra Sound is said to have been the site of the last surviving community of Norsemen in the Barra district.

Next to none of this information, it need scarcely be said, was derived from printed books, still less from the formal compulsory school education given on the island, which was entirely in English from its inception in 1872 until 1918, when Gaelic was admitted as a permissive subject, and where the teaching of history was heavily coloured with the pro-Whig and anti-Highland bias of the standard Scottish text-books. The vehicle of Barra tradition was the Gaelic folk-tale, the anecdote, and the folk-song, of which thousands existed; a rich and varied tradition, but one which, lacking patrons, brought its practitioners no material reward. Here it was greatly to the Coddy's credit that, unlike some Gaelic-speaking Highlanders who have made their way in the world, he never turned his back upon the language and traditions of the race to which he belonged, but was fully aware of their beauty and did his utmost to encourage those who were trying to preserve these things and prevent them from falling into oblivion. In this my friends and I have been heavily indebted to him.

In 1933 there were living within two miles or so of Coddy's house Ruairi Iain Bhàin and his sister Bean Shomhairle Bhig, the two most outstanding folk-singers I have ever listened to; Seumas Iain Ghunnairigh, an excellent story-teller; Murchadh an Eilein, born and brought up on the now uninhabited island of Hellisay, and full of interesting stories and local traditions; Alasdair Aonghais Mhóir, a famous character and gifted raconteur; Neil Sinclair 'An Sgoileir Ruadh' Schoolmaster at Northbay, descended from Duncan Sinclair who lived on Barra Head,[1] a beautiful speaker of Gaelic who took a most intelligent interest in his native language and aided many of the Gaelic students who visited Barra; and many others, including Miss Annie Johnston, well known to many folklorists, who lives in Castlebay. Doyen of all these was Fr John MacMillan, parish priest of Northbay, a native of Barra, great in heart and in body, a wonderful preacher in Gaelic, and a true poet. No student of Gaelic could wish for better surroundings and company. Looking back on those days my great regret is that we had not the means to record these tradition-bearers adequately before the Second World War broke out. In those days such work was entirely unrecognised in Scodand, though this was not the case in other countries such as Ireland, Scandinavia and America. I remember very well trying, in 1938, to find an institution in Scotland which could accept copies of recordings of traditional Gaelic songs made in Nova Scotia the preceding year. I could find none which had any provisions for accepting such a gift.

The process of getting inside the tradition itself was by no means easy. First the local dialect had to be learnt; here 'book Gaelic' was an actual obstacle. All spoken Scottish Gaelic dialects differ from

---

[1] See *Gairm*, Vol. II, p. 271.

the literary language, in some respects consistently: the dialects of the Outer Hebrides are, in fact, more vigorous than the modern literary language, and contain many words and expressions that are not in the printed dictionaries.[1]

Coddy was assiduous in assisting these studies. We used to go together to the houses of Seumas Iain Ghunnairigh or Alasdair Aonghais Mhóir in the winter evenings, when the story-telling and exchange of reminiscences would soon begin. At first I could hardly do more than pick out an occasional word or sentence here and there. I was just beginning to feel that I would never do more than this when suddenly things seemed to become clear, although, of course, there were (and are) still many difficulties to be overcome.

In January 1937, much to the Coddy's interest, I was able to acquire a clockwork Ediphone, after having first tried and discarded a perfectly useless phonograph recommended by some foreign authority. This Ediphone interested the Coddy hugely and he was assiduous in finding people to record on it. In particular, he was anxious for the songs of Ruairi Iain Bhàin to be recorded: for Ruairi was getting old; his sister Bean Shombairle Bhig, a wonderful folk-singer, who had sung, to Mrs. Kennedy Fraser, was in ill health and unable to sing again.

In July 1937 my wife and I, who had made our home in one of Coddy's houses since 1935, left with the Ediphone to record old Gaelic songs from the descendants of Barra emigrants in Cape Breton,[2] amongst whom we found the same songs still preserved by the old people; that is another story. On our return to Barra in 1938 I brought a Presto J disc recorder with which I was in time to re-record Ruairi Iain Bhàin and a number of other singers; some of these recordings with a book of words were published by the Linguaphone Institute in 1950, greatly to the Coddy's joy. When I visited Barra again between 1949 and 1951 with a wire recorder, Coddy was again to the fore in encouraging the work, and recorded Gaelic versions of twelve of the anecdotes which are printed here. He realised perfectly well what so few realise: that the Gaelic oral tradition is an immense tradition containing many things of beauty

[1] It is hoped that the publication of the collection of these compiled in South Uist by the late Fr Allan McDonald sixty years ago, by the Institute for Advanced Studies in Dublin, will stimulate the study of Hebridean Gaelic and make the work of Hebridean poets more easily comprehensible. [2] See 'A Visit to Cape Breton,' *Scots Magazine,* September and October 1938.

and interest, and a great deal of it had already perished irrevocably, and that unless a desperate effort were made, most of the rest would perish too, especially the old songs *in their authentic form*. 'Ah, Iain,' he used to say, 'if only you had come around with that machine twenty years ago, what wouldn't you have got from the men and women who have passed away. You could never believe how much has been forgotten.'

In the spring of 1951 I made my last visit to the Isle of Barra with the wire recorder, accompanied by Mr Francis Collinson, who had just been given a Research Fellowship in folk-music at Edinburgh University. As usual, the Coddy was to the fore in encouraging the work and, of course, we stayed at his house at Northbay, where another visitor was Miss Sheila J. Lockett, on a holiday from London. After hearing some of the Coddy's tales, Miss Lockett was so much struck by their interest and the vividness of the Coddy's style, that she suggested they would be well worth taking down in shorthand. This was eventually arranged, and later in 1951 and in 1952 Miss Lockett made a number of visits to Northbay for this purpose, taking the stories down in the Coddy's own words, just in time before his memory began to fail. Miss Lockett, whose name should be known to students of Scottish Gaelic folklore in connection with the part she has played in drawing the music which is printed in *Folksongs and Folklore of South Uist*, and in preparing this work and Fr Allan McDonald's *Gaelic Words from South Uist* for the press.

A word on the history of Barra will not be out of place here, for it is a constant background to these stories. Traditionally the name of the island is associated with St Finbarr of Cork who lived in the sixth century. St Finbarr, whose story can be read in Charles Plummer's *Lives of the Irish Saints*, was undoubtedly a powerful saint, but there is no record of his ever having been in Scotland. More probably the church was dedicated to him by one of his pupils. At any rate, the memory of St Barr is still vivid in Barra. Down to the seventeenth century it was popularly believed that dust from the burial ground named after him at Eoligarry, scattered on the sea, would result in the calming of storms, and even later his statue was preserved in the church there, as is mentioned by Fr Cornelius Ward (1625) and Martin Martin (1690). This statue, indeed, may still exist somewhere in concealment.

In the ninth century came the Norsemen, who left their mark in the form of Viking burials, popular traditions and place-names like 'Breibhig' (Broad Bay),' 'Alla'asdail' (Elves' Milking-place), 'Hellisay' (Cave Island) and so on. *Grettissaga* tells how, around 871, the Viking Onund expelled from Barra the Irish King Cearbhall (Carroll). From that time on the Norsemen retained possession until the Kingdom of the Hebrides was re-Gaelicised under the Lords of the Isles from the thirteenth century onwards. Local traditions point to Norse-speaking communities having survived on the small neighbouring islands, the last such being on Fuday, which was extirpated by 'Mac an Amharuis.'[1] Viking galleys continued in use amongst the islanders until suppressed by the Privy Council of Scotland in the early seventeenth century as a means of disarming the local Hebridean chiefs who had risen to power following the breakdown of the semi-independent Lordship or Kingdom of the Isles at the end of the fifteenth century.

During these times the MacNeils of Barra, who at one time appear to have had ambitions in the direction of annexing the Isle of Coll, were allies of the MacLeans; but from the seventeenth century onwards, they became associated with the MacDonalds of Clanranald, their neighbours in South Uist.

This Lordship, which probably gave the islands the best (and only native) government they ever had, was an institution very much valued by the Hebrideans, who made several attempts to restore it during two generations after its suppression. Finally the islands lapsed into a barbarous anarchy during the second half of the sixteenth century, with third parties egging on internecine clan warfare in the hope of ultimately benefiting, a strife which was worsened by the collapse of all religious sanctions in the Isles following the Scottish Reformation of 1560. From this state of affairs emerged an official Government policy towards the Isles that was almost entirely repressive so far as the language, religion and traditional institutions of the Islemen were concerned.

During the time of the Lords of the Isles the first historical reference to the famous MacNeils of Barra occurs. This was in 1427, when the Lords of the Isles, who of course were MacDonalds,

---

[1] The late Dr George Henderson encountered this tradition over 50 years ago. See his *Norse Influence on Celtic Scotland*, pp. 42, 177. See also J. L. Campbell, *Sia Sgialachdan*, p. 7.

gave a charter of Barra to 'Gilleownan Roderici Murchardi Macneil,' *i.e.* Gille Eóghanain mac Ruairi 'ic Mhurchaidh 'ic Néill (Gilleonan son of Roderick son of Murdo son of Niall). This is the reason why a seventeenth-century South Uist poetess flyting with the Barra woman Nic Iain Fhinn, refers to 'little stony Barra' as:

> 'Deirc a fhuair sibh bhuainn an asgaidh
> Nuair a chunnaic sinn nur n-airc sibh.'

'Alms which you got from us [*i.e.* the MacDonalds] for nothing, when we saw you in want.'

Reckoned from the Niall mentioned in this charter, General Roderick MacNeil of Barra, last male chief of the direct line, who died in 1863, was the sixteenth of the line. The Clan genealogists, however, did better than that: according to Martin Martin, the Chief who lived at the end of the seventeenth century was accounted the thirty-fourth of lineal descent who had possessed the island. This descent was linked up, by the Gaelic shenachies, with the Uí Néill of Ulster, including the famous Kings Niall of the Nine Hostages and Niall Frasach (so called because of several extraordinary showers which fell in his reign). Of course, there is no historical proof of this descent, and the MacNeils of Barra may well have been originally a Norse family which later Gaelicised its family tree, but in any case, the family was sufficiently old and distinguished, and the Coddy was always proud of his connection with the Clan MacNeil through his paternal grandmother Mary MacNeil, daughter of Neil MacNeil of Greian.

Even in historical times there are difficulties over the MacNeil of Barra pedigree, and these difficulties are not lessened by the total disappearance of the family papers, including the famous Barra Register, which was a Gaelic Chronicle of the family similar to the Red and Black Books of Clanranald. It is possible that these papers still exist somewhere: it is known that the Barra Register was in the possession of members of the MacNeil of Vatersay family towards the end of the nineteenth century. This family was a collateral branch of the Chiefs' house, and its original progenitor may well have been Niall Uibhisteach, the rightful heir of the estate who was dispossessed by Niall Òg in 1613, and forced, as Fr Cornelius Ward records in 1625, to sign away his lawful rights in Niall Òg's favour.

Niall Uibhisteach and Niall Òg were both sons of Ruairi an Tartair ('Noisy Rory'), a famous old warrior who flourished at the end of the sixteenth and beginning of the seventeenth centuries. Apparently Niall Uibhisteach was the son of a legitimate marriage with a daughter of MacDonald of Clanranald (as his name implies), and Niall Òg was the son of a liaison with Mary MacLeod heiress of Dunvegan and widow of Campbell of Auchinbreck.[1] There was a good deal of strife between the sons of the two mothers (as occurred in other similar cases in the Isles), and a complaint was made to the Privy Council of Scotland in 1613 that Niall Òg had imprisoned his father and half-brother and had seized control of the estate. This situation was apparently never rectified.

Ruairi an Tartair referred to was a famous pirate and is said to have instituted a search for Rocabarraidh (*i.e.* Rockall) in the belief that it was inhabited by men who owed him allegiance, and to have raided the West Coast of Ireland. During his time the Barramen continued to make religious pilgrimages to Croagh Patrick in Mayo, and when the Irish Franciscan Fr Cornelius Ward visited Barra in 1625 at the invitation of one of the local gentry, he reconciled a large number of persons without difficulty, the usurping chief excepted. He in due time was reconciled by Fr Hegarty in 1632, and the family remained staunch Catholics until after the '45 – the clansmen ever since, except one or two families. In the 1650's the famous Vincentian Fr Dugan, who is still well remembered in local tradition, made similar visits. There is thus no foundation for the story told to Martin Martin around 1690 by the minister of Harris, that the Barramen had been Protestants before the Restoration of the Stuarts in 1660.

To return to Ruairi an Tartair. It had been frequently said that owing to a piracy he had committed on a ship of Queen Elizabeth of England, Roderick MacKenzie, Tutor of Kintail, a well-known Highland diplomat of those times was commissioned to apprehend him.[2] This was carried out by the aid of wine, and Ruairi was kidnapped and taken to Edinburgh and brought before King James VI and I, who asked him (through MacKenzie) why he had committed such a crime. To which Ruairi replied that he was only avenging King James' mother, Mary Queen of Scots, whom Elizabeth had

---

[1] See the article in Vol. V of the *Innes Review,* p. 33, where this question is discussed in detail. [2] See the Coddy's version of this story, p. 42.

judicially murdered. Taken aback by this reply, King James allowed him to go free, on condition of accepting the Tutor of Kintail as his feudal superior.

The drawback to this story is that it is chronologically impossible. Roderick MacKenzie became Tutor of Kintail in 1611, and the only time King James visited Edinburgh after succeeding to the English crown in 1603 was in 1617, by when Ruairi an Tartair was a very old man languishing in irons at the hands of his unfilial and illegitimate son Niall Òg. However, there certainly is foundation for the story. The MacKenzie chroniclers mention that the Tutor of Kintail had dealt with a recalcitrant MacNeil of Barra; but they do not say *which* MacNeil. Indisputably the Tutor of Kintail got a Charter of Barra in 1621 'for services rendered.' I incline to the belief that Niall Òg was the person he really brought to heel.

An interesting and well-authenticated story of the MacNeils of Barra which has, curiously, not survived in local tradition[1] as far as I have been able to discover, is that of the occasion when John Mac-Leod of Dunvegan sent a boat containing a King's Messenger and several men to collect a debt of 85 merks (£56 13s. 8d. Scots or £4 14s. 5d. sterling) from Roderick MacNeil of Barra in 1675. When the party arrived at Kismul they found the gate locked against them and were greeted with stones and shots from the battlements. Driven off, they took refuge on Fuday, where MacNeil's men, headed by his brother James – a person who is not mentioned in the MacNeil clan histories – and Iain mac Néill Uibhistich (presumably a son of the dispossessed heir of 1613), Baillie of Barra, pursued them, hid their oars, caught the King's Messenger and deforced him of his summons, and then released him, telling him never to show himself in Barra again.

Deforcing a King's Messenger was a serious offence, and Mac-Neil, his brother James and various other persons alleged to be in-volved, were tried in Edinburgh in 1679 for it. Owing to lack of evidence, MacNeil, who was alleged to have been seen in Kismul Castle in disguise watching the King's Messenger being driven off, was acquitted, but his brother James was sentenced to be fined £1000 (Scots) of which 500 merks were to be paid to the clerk of the court for the King's use, and 1000 merks to be paid to John

---

[1] A condensation of the official report is printed, unfortunately with some gross inac-curacies, in the *Book of Dunvegan*, I, 180–2.

MacLeod of Dunvegan, and to be imprisoned until these sums were paid, as they presumably were by the chief. It must be said that the fact that MacLeod went to the length of employing a King's Messenger to collect a very small debt from MacNeil of Barra indicates that there must have been bad feeling between the two already, so that more probably lies behind this story than meets the eye. An interesting allusion in the official account of the case is to Alexander Shaw (? Shein), schoolmaster. There was no official school on Barra in 1675 as far as is known, and Shaw may have been a 'hedge schoolmaster' on this Catholic island. In 1703 the names of Donald and Angus Shaw are found in a list of the leading Barra Catholics, along with those of the Chief and his five children and two brothers, and the MacNeils of Vatersay, Tangusdale, Greian (who may be the ancestors of Coddy's grandmother) and Vaslin.

From 1621 until it was redeemed by General MacNeil shortly before the sale of Barra in 1838, the feudal superiority of Barra was held – apart from a brief interlude in the short reign of James VII and II – by, firstly, the MacKenzies and secondly, the MacDonalds of Sleat, the annual duty paid by the MacNeils being £40 and a hawk. In the event of the MacNeils being forfeited for, *e.g.* rebellion, the estate would have reverted to their feudal superiors. This very nearly happened in 1746 owing to the intrigues of the then MacNeil with the Jacobites, and his open training of men to join Prince Charles' army. The Chiefs and Clan were always Jacobites, and fought at Killiecrankie and at Sheriffmuir for this cause, and would undoubtedly have joined Prince Charles in 1745 had the Jacobite army not got so far away on the mainland. As it was, arms and money were landed for the Jacobites on Barra, and the then Chief trained men openly in the winter of 1745–46 with the hope of joining the Prince's army. All this was known to the authorities, and when receipts signed by MacNeil of Barra were found on a Spanish agent captured in 1746, no further evidence was needed to justify his arrest. He was taken to London along with other prisoners but later on was released. It has been said he turned King's Evidence, but research in the State Papers Domestic has not confirmed this allegation so far.

After the '45, it seems that MacNeil of Barra's heir, Roderick, abandoned the Catholic religion and obtained a commission in the army (for which Catholics were then ineligible) and was killed at the siege of Quebec (see p. 82). The Vatersay family had already become Protestant around 1727, after a second marriage to a MacLeod of Greshornish, and had intrigued thereafter with the SPCK for the foundation of a Protestant school on Barra – a thing which could only have been embarrassing and irritating to their Chief. Nevertheless, they were intriguing with the Jacobites in 1745. The Vatersay family obviously possessed a good deal of influence with the chiefs in the eighteenth century and may have thought, at one time, of claiming the estate as Protestant heirs, under the Penal Laws.

Roderick MacNeil of Barra succeeded as owner and chief in 1763 and lived until 1822. He was thus laird for practically his whole life. During his time emigration to North America started and for a time was much opposed by the estate, as the people, whose labour was needed for the kelp industry, were being deluded by ridiculously rosy pictures of life across the Atlantic drawn by emigration agents who had a financial interest in obtaining as many passengers as possible. Kelp provided big incomes for island landlords at this time, whose power was so great that they were even able to forbid the local fishermen from trading their fish direct with Glasgow or Greenock merchants.

This Roderick MacNeil married Jean, daughter of Cameron of Fassiefern, and had two sons and five daughters. In 1806 he drew up a Deed of Entail designed to keep Barra in his family for ever, and to prevent debts and family settlements from accumulating on the estate. This Deed names as his heirs his sons and their male heirs, whom failing his daughters and their male heirs, whom failing his sister and her male heirs, whom failing Roderick MacNeil, tacksman of Brevig. This Roderick of Brevig is the ancestor of the late Wallace MacNeil of Vernon River, Prince Edward Island, and of Colin MacNeil, his brother; the family claim to be descended from Gilleonan, the second son (or the first son passed over) of Roderick MacNeil of Barra who flourished *c.* 1688. The MacNeils of Earsary, whose representative, Mr Robert Lister MacNeil, now owns Kismul Castle and part of the mainland of Barra, assert, however, that the Brevig family is descended from an earlier generation and that their (the Earsary) family derives from James, second (of only two) sons of the Roderick MacNeil of Barra referred to. This debate between the Brevig and Earsary families ignores, it may

be said, the claims of the Segraves, the descendants of General Mac-
Neil of Barra's only child Caroline (the heirs of line) and of the senior
male descendants of his sisters, the Campbells of Achallader (who
may be considered heirs male of Entail), so that, given all the available evidence, the professional genealogists have quite a pretty knot to
disentangle regarding the appropriate arms for the different descendants of this ancient race.

Kismul Castle, it may be said, was abandoned as an inhabited
dwelling around 1700. Generations earlier, a seer had foretold how it
would one day become the home of otters and seabirds, thereby earn-
ing the displeasure of the then MacNeil of Barra. George Wilson,
who visited Barra in 1841, on a voyage around the coasts of Scotland,
says that an old woman living in Barra was named to him as a person
whose mother had been born within the Castle. The Castle dates, according to the Ancient Monuments Commission, from the fifteenth
century.[1] After its abandonment, the MacNeils of Barra eventually
built the Mansion and walled garden at Eoligarry. There are interesting allusions in some of the Coddy's stories to life in Kismul.

Colonel Roderick MacNeil of Barra died in 1822. His will, made
in 1820, together with the Deed of Entail executed in 1806, tied
the estate up very tightly; apparently he was not confident in the
financial discretion of his heir, Roderick MacNeil, who had fought
at Waterloo, and later became a general. After making elaborate
provision for management of the estate by trustees in order to reduce the burden of debt on it, Colonel MacNeil burdened it afresh
with large settlements in favour of his younger children. His suc-
cessor referred to himself as having been literally tied to the stake;[2]
and when the price of kelp failed he was put in an impossible situation. Unhappily, guided by the advice of persons ignorant of the
country, he resorted to one money-making scheme after another,
organising fisheries, starting a glass factory at Northbay and im-
pounding his tenants' cattle – one agent or factor succeeded another on the estate until things got into inextricable confusion and
Barra had to be disentailed and sold to pay off his creditors, including presumably his younger brother and sisters, in 1838. Hot

[1] The Report points out that it is not mentioned in Fordun's *List of Castles in the Isles*,
compiled about 1380. [2] In a letter to Fr Angus MacDonald, dated 27th October 1823.
See the *Book of Barra*, p. 173.

tempered and tyrannical as General MacNeil of Barra appears to have been, he is remembered with affection locally by comparison with the absentee régime, associated with rack-renting, evictions and all the petty tyrannies of unsupervised subordinates, that followed him. The years of 1822 to 1886 were, indeed, a time of poverty and oppression in Barra, as in the Highlands and Islands generally. This was relieved by the passing of the Crofters Act in 1886, granting fair rents, security of tenure and compensation for improvements. No one can nowadays visualise the petty tyrannies that were possible in the days when Island estates were governed by the representatives of absentee landlords and the people were completely without these indispensable rights and without a democratic franchise.

As has been said, the Coddy's maternal grandfather, Robert Mac-Lachlan, came to Eoligarry from Mull as gardener in the time of General MacNeil, and his paternal grandfather, Angus MacPherson, came to Craigstone from South Uist in 1819 to do joinery work there for the priest of Barra, Fr Angus MacDonald. Thereafter the fortunes of his family were bound up with those of Barra, which suffered severely from the ruin of the local inshore fishery which followed the coming of the steam trawlers from Fleetwood around the turn of the century, against which the under-policed three-mile limit was a completely inadequate protection (in Norway, for example, waters like those of the Minch are totally closed to trawlers). In more recent times this loss has been followed by the equally disastrous closing of the herring curing stations on Barra. The policy of Westminster Governments appears to be to concentrate life in the Scottish islands in a few urban centres and let the outlying districts die.

Sea communications with Barra remained so inadequate and uncomfortable down to 1929 that very few strangers had visited the island before that time. This, no doubt, helps to account for the preservation of its Gaelic traditions, already referred to. After the improvement in these communications, Coddy took full part in the development of the tourist traffic to the island.

Today, with the failure of the fishing, the lack of local industry and with the growth of a civilisation which cannot easily provide rural communities with the amenities enjoyed by towns, Barra is left to rely on pastoral agriculture, the tourist traffic (for which the season is brief) and some lobstering and cockle gathering – and even the lobsters are encroached on by large boats from the mainland. It is a sad future for the community to which the Coddy belonged and for which he did so much service. Perhaps things might have been better had there been a different disposition towards the remote parts of Scotland at headquarters. It is equally possible that, but for the sympathy for and interest in Barra that the Coddy's personality was able to arouse amongst strangers, things might have been considerably worse. At any rate, nothing can rob Barra of its beauty, or of the memory of its splendid tradition of folk-song and story; and many of us have to thank the Coddy for an introduction to all of these. We may therefore wish that his memory be long preserved.

J. L. CAMPBELL

Isle of Canna,
22/1/59

Tales of the MacNeils of Barra
and other Lairds

The family tree of the MacNeils

Roderick son of Roderick son of Roderick son of Roderick son of
Roderick son of Roderick son of Roderick, chiefs of Barra. The
first Roderick was a son of Gill' Eóghanain. It happens that Gill'
Eóghanain was a son of Ruairidh, and Ruairidh was a son of Murchadh son of Niall the Fair, son of the King of Ireland (Niall bàn mac
Rìgh Éirinn). And the River Nile was called after him as he owned
territories all round about the Nile.

The seventh Roderick, proprietor of the Island of Barra, fought
with much distinction on the field of Waterloo in 1815. In 1838 he
went bankrupt and sold the island of his forefathers to Colonel John
Gordon of Cluny, Aberdeenshire.

[*See the account of the MacNeils of Barra in the Introduction. This
genealogy is telescoped. It seems that the last Roderick was really the sixth.
See Pedigrees, p. 13. The reference to the Nile recalls Niul, son of Fenius
Farsaidh, husband of Scota, the mythological ancestor of the Gaels.*]

MacNeil who fought at the battle of Bannockburn

Traditional story on the Island of Barra of how the MacNeils went to
Gigha and Kintyre

In 1314 at the battle of Bannockburn it so happened that one of
the famous clan MacNeil was present. Before he went into battle,
while preparing for it, he drew the attention of Robert the Bruce
– he stripped down to his kilt only and his braces – that is all. Now
the battle began and Bruce kept his eye on this man the whole time
– or practically so. He noticed that MacNeil was making a mark of
distinction on the field with his battleaxe and that he was mowing
the English enemy down wholesale and retail. Now King Robert
was very very satisfied – not only with MacNeil but with all his men

– and very early in the day he said to himself that the victory was in his hand. After the day was over Bruce gave an order to one of his officers to call this man specially to see him, along with a few others, and they were brought along into a hall. Those were the people who made the most distinction on the battlefield. Calling MacNeil to his presence he asked him what part did he come from. And MacNeil replied that he came from the Island of Barra. 'Ah, yes. Well,' he says, 'I am thanking you very much for your bravery today and,' he says, 'I shall have an interview with you later.' he says, 'after I see and speak to several others who fought so brilliantly with you.'

Then King Robert went round them all, and when he had interviewed them all he called MacNeil a second time. And he says, 'Now, MacNeil, I am going to ask you, can you get, say, twenty men – good men like yourself – and take them with you out to the other islands and the mainland of Argyll?' And MacNeil said, 'Yes, your majesty, I will be pleased to do that.' So one day MacNeil and his party left the island of their forefathers to take up their abode in islands and places that were strange to them. On arriving, King Robert interviewed them and divided them among the islands of Gigha, Colonsay, Juray, Islay and Mull and the county of Argyll. And that is how the MacNeils went out from Barra and increased and multiplied in the islands I have mentioned.

The descendants of the men who left Barra came up generation by generation, being of the same build as their forefathers, daring seamen, plucky men and fearless, and from those islanders came people who went all over the world, masters of sailing ships, and explorers of various countries, with the result that you will today find MacNeils in every corner of the universe, and each and every one of them making a mark of distinction in life.

[*This story implies that the MacNeils of Barra are senior to the Mac-Neills of Gigha and Taynish, which is denied by the latter. The connection between the two clans, if any, is a very remote one.*]

MacNeil's raiding of Iona

Once upon a time the MacNeils went on a raid to Morvern on the north side of the Sound of Mull. They walked the bigger part of the night right along in that direction and got nothing. Whether, now, the inhabitants knew they were coming or not is not clear, but there was no cattle to be seen in any direction. They walked right up until they came opposite Iona. And, of course, the galley was following up the Sound the whole time. And then they decided to cross the Sound of Mull and go on to the Island of Iona. That was about in the very dead of the night, or maybe in the early hours of the morning, and they did not see anything. And then they went into a churchyard. All was still and silent, nothing could be seen except the remains – which they could not see!

Then they had a conference and said to themselves that it was not very good to be going back to Barra without anything, and they had no cattle, and so, in the absence of cattle, they decided to take nine stones of those that were in the churchyard – nine stones, rather than go back to Barra empty-handed. And the motion was carried and each and every one of the nine took a stone with him. And every one that took a stone was a MacNeil, and each and every one of the stones were put on the graves of MacNeils in the churchyard of Kilbar. And it so happens that one of them is on my own grandmother's, Mary MacNeil, and her father's name was Neil MacNeil, and MacNeil my great-grandfather, he was a prisoner of war in the Napoleonic Wars in France. I will give you a separate story about that.

On the way home, and crossing the Minch, they met fearful weather, and each and every one had to take an oar and row across the unusual hurricane. Now MacNeil expected the Kismul galley would be crossing the Minch that day. So he sent the fiery cross round the island for everybody to keep a look-out on the Minch, and especially on the top of Beinn a' Charnain, that is a hill on the east side of Castlebay.

Now there were very many of the islanders kneeling and praying for the safety of the famous galley. One man, better than many others on the hill, saw what he thought was spray going over a boat and he told them to look out in that direction. They made out that it was the galley coming in this terrific weather, coming nearer and nearer, and latterly they could make out that the men were sitting on their oars, stripped to their shirts, rowing against that terrible hurricane. And MacNeil sent a message to the castle to prepare a feed of the bullock that was hanging in the castle for their arrival. But in further orders, they had not to get any food or drink for the next twelve hours, and the further statement that the beds be prepared for them and the blankets heated up that were spun and woven on the island.

And latterly they came into the bay, and so great was their exhaustion that some of them had to be carried out from the shore to the castle. MacNeil summoned four men to rub them down with whisky, the oldest stock in the castle, but to be sure at the same time not to give them any to drink. Now the operation began and the men were carried ashore, and as they were coming ashore there were two men rubbing them down, rubbing the whisky well into them from the top of their heads right down to the soles of their feet, and strict orders were given not to give them anything to eat. They rested and slept for twelve hours and some of them were that badly bruised they could hardly turn from one side to another, with the muscles strained almost out of existence. At the same time the *gocman*[1] gave orders to the butcher in the castle to cut up chunks of the bullock that was hanging up in the castle, and the fire brought out and spread, and chunks to be roasted on the fire – there were no ovens or anything like that.

Now when the twelve hours were up, he gave orders to call them and he saw that they got an equal division out of the cask, the oldest in the castle, and taking particularly good care that no man could or would get any quantity that would do him harm. Sitting round and telling each other news about the hurricane and news about Iona, and news about the stones they were going to put at Eoligarry kept them talking. He further told them to rest for a few days. And the boys, after two or three days' rest sent a message to MacNeil that if he wanted them to go out on another raid that they would go. And he cautioned them on the ground that they were so tired they had better rest for a few days more. And before they did go anywhere, each and every one took the stones and put them in their own burying ground at Eoligarry.

One of the stones was removed to Edinburgh for identification. There was nobody who could identify it and so two of them were sent to Denmark. And there they were identified and sent back to Edinburgh, and from Edinburgh to Eoligarry.

[*If the stones were taken in this way it was presumably for ballast. The incident could have occurred in the sixteenth century when the MacNeils were allied with the MacLeans of Mull against the MacDonalds. The*

[1] The look-out man at Kismul Castle.

stone referred to in the last paragraph is probably the rune-inscribed stone that was taken to Edinburgh.

This story was recorded by the Coddy in Gaelic on 3/1/50 and in both Gaelic and English on 3/4/51.]

MacNeil of Barra, the widow's son and the Shetland buck

MacNeil one day went to the Island of Mingulay. He met a woman on the island who was a widow and had only one son. He wanted to take the son home to the castle where he would be well looked after, and he gave the mother to understand that he would never want, and that he MacNeil, would be very good to herself. With tears in her eyes she said a more difficult thing than to part with her boy she did not know. At last he said, 'I will be taking him over to see you occasionally when I visit the island collecting rents and so on.' And latterly when she learnt there was hopes of seeing him occasionally, she was very pleased, and on these conditions she did agree to give him the boy. And this went on for several years; MacNeil did as he promised – he would be taking the boy over on occasions of his own visits to the island.

Now MacNeil was fond of wrestling, and when the young boy grew up in strength MacNeil used to practise wrestling with him. And that went on for a period until one day the widow's son caught hold of MacNeil and gave him a firm grip which compelled MacNeil to go flat on the broad of his back on the ground. Everything went well until it reached that stage. After that day MacNeil took into his head that the day was coming when he would be beaten, and so he made up his mind to kill him and another man that was boatman in the *birlinn.* So when the time was ripe and the weather was there to carry out MacNeil's intention, he gave the *gocman* the order to sound the trumpet, and called the *birlinn's* crew. At the moment we are talking about it was blowing a hurricane – MacNeil was doing this trick simply to drown the man in charge of the *birlinn* and the widow's son.

Now the *birlinn* was brought alongside, the widow's son was standing in the bow and he gave them all a hand to go on board. It came now to the stage that McNeil was going to bid them good-bye and he did so by shaking hands with the bowman. He caught MacNeil by the wrist and instead of shaking hands with him he flung him down amidships, and then he gave the order to shove off.

Now MacNeil was very silent. He was not speaking at all, and when they went past the castle he said, 'Well, boys, the weather is not suitable for the job – we'd better turn back.'

'We will not turn back, but we will keep on. It was good enough for us to go, but when it came to you the weather was bad!' said the widow's son. So on those grounds there was no more about it but to carry on – how far we don't know.

Now following out, the wind was in the nor'-west, and they decided to go to the back of Muldonaich at the foot of Sloe na h-Iolaire. They were there the whole night keeping her stem right on to the rocks, but the squalls and the gusts and the spindrift was going high on the outside of them. It carried on all the night until the break of day and then the wind abated. Then it cleared up and the day was very good. They all held a conference and decided not to go further than that for this time – they were very exhausted, to start a raid into which they needed to go very fresh. So the raid was put off, they came back to the castle and the usual performance took place – to have a big dinner and plenty to drink and to eat, and wait for a more suitable opportunity.

The next visit to Mingulay the widow's son went over and got an opportunity to see his mother. Now he told her the story – how Mac-Neil did attempt to drown him and drown another man in the *birlinn*. She knew the other man herself, and she was very sorry to hear it. 'But,' she says, 'it is very difficult for me to take you from MacNeil now.' 'Mother,' he says, 'the time is not very far away when I shall be going anyway, and that is the second time he has made an attempt on my life. But I am very sure he is not going to get another opportunity.'

So the widow's son bid goodbye that day, and more than probable he never saw her any more.

[*This is the version taken down by Miss Lockett: the remainder is transcribed and translated from the Gaelic wire recording made on 1/1/50, and printed here as an Appendix.*]

Now, time was passing, and MacNeil went and challenged every man in Scotland and outside, in the Isles, in Shetland, and everywhere around – that a challenge would go out for someone to fight against him. He got a challenge, and it was from Shetland he got it, from a famous man there whose like was never on his land, called the Shetland Buck. He sent out the challenge, 'I'll take you on,' he said.

Now, the day came that had been set apart for the fight, and they met in Shetland. MacNeil was frightened enough at going and he said to himself, 'I'd better take John, the widow's son, with me, in case he (the Buck) does me any harm.'

So John went with him. They arrived in Shetland, and MacNeil said, when he saw him (the Buck) coming, 'Now,' said he to John, 'if you see me putting my hand behind me twice, you'll understand it's time for you to intervene, and to come between me and the Shetland Buck.'

But without making a long story short, or a short story long, the fight began, and a fierce fight at that, and at last the widow's son saw MacNeil put out his hand, and he watched and saw him do it again. He went in and stood between the Buck and MacNeil and said to him, 'I'll support you' – and then began a contest and a battle that was worth calling a battle. The widow's son didn't take long to put the other hero on his back, and he put his knee on his chest. The Buck asked for mercy, to let him up and they would make peace.

'I'll do that,' said John. 'I'll let you get up undoubtedly if you admit you're beaten.'

'That's not difficult, everyone sees I'm beaten.'

He let the Buck get up, and when they were talking, who jumped over but MacNeil, and drew his *sgian dubh* (black knife) from its sheath and was going to stab him. The widow's son turned to him and said, 'If you put a knife, black or white, in the man, I'll take care right away that you won't see Barra again in your life.' So it was. Mac-Neil took fright and abandoned the matter entirely.

The Buck and the widow's son shook hands, and the Buck said to him: 'You will be with me in Shetland all your life, and any descend-ant you may have; and you may marry whoever you wish, and you will be rich and happy and quiet along with me in this world. And –' he said to MacNeil, 'you can go home, MacNeil; you shouldn't have come here at all since you didn't know how to fight like this good fellow who was the first ever to put me on the ground,' he said.

The widow's son remained in Shetland along with the Shetland Buck, and down till today you find MacNeils there descended from him.

MacNeil of Barra and MacKenzie of Kintail

Once upon a time MacNeil of Barra was besieged by the MacKenzies of Kintail, and a big lot of the MacKenzies came to Barra. And the target of their mission was to starve MacNeil in the castle, and as they came very unexpectedly and MacNeil was without any preparations to meet the besiegers, very soon MacNeil said to himself, 'This man is going to do me out.'

One night he was sitting by the fireside in the castle and he began to ponder how to get out of the difficulty. The sentry was so dutiful it was impossible to escape out of the castle. And as he was pondering, the thought came to him, and this is what he decided upon. He had two dogs in the kitchen of the castle lying down and he said to himself, 'Well, I will bleed the two dogs to death and there is a pile of sheepskins in the castle, and I will paint the insides of the sheepskins with the blood of the dogs and put the skins on the parapet of the castle and then MacKenzie would say to himself, "I may as well clear out – if MacNeil can kill all these sheep I shall never starve him out."'

Now he got hold of the dogs and bled them one by one. And the first one he killed he smeared the blood inside the skins and as he was doing so he was putting the skins on the parapet. And then he killed the second dog and dealt with it likewise, smeared the blood on the rest of the skins and hung the skins on the parapet of the castle. So when MacKenzie's sentry next morning saw the position he told it to his commanding officer that there was no use to wait until he had starved MacNeil out of the castle – that in his opinion MacKenzie would starve first. So they had a conference and they came to a decision that the wisest plan was to clear out.

★ ★ ★

Several years passed and MacKenzie discovered how MacNeil did him in the eye, and he was very furious and he said, 'Well,' he said, 'I will have my own out of MacNeil one day.' Now MacKenzie had a ship and went down the coast of Portugal and the coast of Spain and he came home with a big cargo of every kind of wine on the calendar. Coming down the Minch he decided that as he was so near MacNeil he would call on him. So he did. And MacNeil invited him in to the castle and they had a little dram there, and when MacKenzie was leaving he said, 'Now MacNeil, you will come out one night to see me,' he says, 'and I have very good wines which I have got in Spain and Portugal and I will give you some of it to take to the castle with you.' MacNeil accepted the invitation very cordially, and next night he went out to the ship to dine with MacKenzie.

The wine was very plentiful and MacKenzie knew the blends which would hit MacNeil the hardest, with the result, I regret to have to tell you, that MacNeil went below the table. Whenever Mac-Kenzie saw MacNeil below the table he gave the order, 'Up anchor!' and the boys stood by and hove the anchor, got the sails in order and cleared out of Castlebay. MacNeil then understood that something was wrong, for the ship was pitching at such a rate that he could not be in Castlebay. And looking up the cabin hatch he could see Barra away in the background, and he just took hold of his venerable beard and, 'Well,' he says, 'I am very disappointed with what happened and we will just see what is further going to happen.' He did not lose his courage although he was beaten.

MacKenzie was proud of the opportunity he had in taking the raider MacNeil to court in Edinburgh. Well, the day came when MacNeil had to stand before the court, and the sheriff knew Mac-Neil and he was also a blood relation of his – although he did not let that on in court.

Now the trial began and the judge put MacNeil on his oath and then he said to him, 'I am going to examine you, Mr MacNeil. Are you the MacNeil of Barra?' says the judge.

'Yes, my lord,' says MacNeil.

'Are you the man,' he said, 'who plundered a ship between Barra Head and Northern Ireland and looted the cargo and took the cargo back to Castlebay?'

'Yes, my lord.'

'And this same crew – you were very kind to them – you gave them plenty to eat and plenty to drink?' said the judge.

'Yes, my lord,' said MacNeil.

Then the next one came. 'Are you the MacNeil of Barra who had the courage to declare war on the King of England single-handed?'

'Yes, my lord,' and MacNeil took a firm hold of the beard and put out his chest with pride and admiration that he was the man who declared war on England single-handed.

Now the judge ceased fire and stopped examining him.

'Now,' he says, 'MacNeil, I cannot say much against you as regards the first offence – it was good of you to save the lives and be kind to the men and take them ashore, but it was very illegal of you to loot the cargo.'

And MacNeil says, 'My lord,' he says, 'looting was quite in order to do in those days, and so I said that I had as much right to loot afloat as ashore.'

'As regards the last part,' he says, 'you declared war on the King of England single-handed. Well,' he says, 'for your courage I will let you go back to Barra without penalising you to any extent. But don't you come back here any more and appear before me in Edinburgh.'

'I will try my best,' says MacNeil, pulling the beard again – and the court was dismissed.

[*See Introduction p. 28*]

MacNeil's return to Barra from the Isle of Man

In the days I am talking about it was customary for the MacNeils to send their children to a college in the Isle of Man, and at the same time every year, about the fifteenth of August[1] the time was when MacNeil went to visit them. So on the twelfth or thirteenth he sounded the trumpet for all the boatmen to muster and get the galley in order to go to the Isle of Man. No sooner said than done – everything was put in readiness and MacNeil gave orders to the *gocman* –'On such and such a date you will send my message to Eoligarry to take up a bullock, and that bullock will be killed at Eoligarry and taken up to the castle, and you will have it in readiness for eating when the galley comes back.'

Well, MacNeil used to take his piper with him on occasions like this, and he took him along this time with him and his name was Donald MacKinnon, and he composed the famous tune called 'Colonel MacNeil's Salute' – *Sealladh nan Ruaridh, Sealladh thog Mulad dhiom.*[2] Everything was ready, and the orders were given to the *gocman* to have the feast ready, when they would return from the Isle of Man.

The MacNeil spent a considerable time along with his two boys

---

[1] The Feast of the Assumption. [2] 'The sight of Roderick, the sight that banished sorrow from me.'

on the Isle of Man, and coming back, between Northern Ireland and Barra Head, there came a heavy fog. Having no compass they were lost. Fortunately it was very calm and they were rowing most of the time. Well, the *creachadair mór* (chief raider) was aboard, and he was an expert at knowing the lie of the land, better than anyone else in the galley. And he noticed the direction the seabirds were flying, and as it was evening he knew they were going to their nests. And they had nothing else to go by but to follow the seabirds.

The *creachadair* went forward and sat in the bow and was keeping a very keen lookout. And now at last he gave a yell. 'Cliffs right ahead,' he said. And shortly afterwards he identified the cliffs and it was the rocks of Barra Head. And then he knew the position beautifully and he said, 'Now,' he says, 'we are all right – here's Barra Head and now we shall follow along the west side, across the sound of Berneray, and right along the shore of Mingulay and across the Sound of Pabbay, and right on till we come to the Sound of Vatersay.' And they followed this course and then they were home.

When they got to the Sound of Vatersay they turned into Castlebay. Now MacNeil knew perfectly well that the *gocman* would be on sentry watch and he advised that they go ashore on the mainland for the night. He knew there was a current there and that they were not far from the castle.

'Well,' he says, 'we had better go ashore, and kindle a fire. I don't want to disturb the *gocman* at such a late hour of the night.' And this they did, and landed at a wee loch there known as Loch Kentangaval, and kindled the fire.

* * *

The bullock was hanging in the castle. The *gocman* noticed there was a little fire over on the shore and he suggested that they take the lungs out of the bullock and put it in the cannon and fire the lungs of the beast ashore and wait and see what would happen. He had no suspicion of enemy attacks, but he was thinking that MacNeil might arrive any time.

Now, here's the cannon leaded and here is the shot fired, which spattered the fire all over the shore. And as soon as this happened MacNeil cried, 'Put out every bit of fire. Or else,' he says, 'If the *gocman* sees another fire, what will happen – you will feel the lead bullets from the cannon in the castle and it will finish you out!' They had to creep there very silently until the next morning. And daylight came and, lo and behold, the beautiful castle was within two hundred yards of them. So they rowed out, and MacNeil went up and met the *gocman* and shook hands with him very very kindly.

'Well,' he says, 'I was very pleased with yourself last night. It shows me what a dutiful, and a very dutiful, servant you are when you saw the light and fired the cannon with the bullock's lungs. And if you had put in the lead or the powder, we would all be done.'

So they went in, and he gave an order to roast a piece of the bullock. There were no roasters in those days – there was a special corner in the castle and the fires were taken out and the bullocks were roasted on iron grids which had to be watched very carefully in case the castle would go on fire with the fat coming out of the beef. And they sat down and had a very good meal.

[*This must have happened before 1700, as the MacNeils ceased to live in Kismul Castle after that date. But I have found no record of MacNeil's children being educated on the Isle of Man. Under the Statutes of Iona (1609) the Hebridean chiefs were forced to send their eldest sons to be educated in the Lowlands of Scotland.*]

MacNeil and the coming of Prince Charlie

Before the '45, it was customary a lot among Highland chiefs to go round their tenants and crofters to see what they could afford to give in support of Bonnie Prince Charlie getting back the throne of Scotland. Money was scarce at the time and funds were very low. MacNeil had a suspicion that one of the crofters of Fuday had a considerable sum of money, and he crossed the Sound one day and went to see Donald MacInnes.

'Well, I suppose,' he says, 'Donald, you have heard that in the very near future Bonnie Prince Charles of the royal Stuart blood is coming back to regain the crown of his ancestors; and,' he says, 'money is scarce and I came over to see you today and see if you could give me some assistance in this direction.'

'Well,' says Donald, a pure Jacobite, 'yes, I will give you money, my chief MacNeil. I will do it without a doubt, especially for the cause. Now, I shall not consider,' he says, 'whether I get it back or not. If I get it back, I shall be so pleased that I don't think I will take it – I shall be that pleased to hear that the Prince got the crown of his forefathers.' And turning round on his heel, he put his hand above the door and from the divots there he took out what is called a *mogan*, that is, the bank they had in those days for keeping their gold.

Well, he counted three hundred pieces of gold to MacNeil of Barra, and MacNeil gave that money to a committee they had in the Highlands for collecting money, and poor MacInnes never saw a penny of it again. After the defeat of Culloden all the gold in Scotland was taken down to London and there were eighteen wagonloads of gold went down from Scotland.

* * *

Now the ship *Ladoutille*[1] came and was piloted to the anchorage by a man called Calum the Piper, who at the time lived in Gigha.[2] (Several years afterwards this same Calum fought with MacNeil of Barra on the Heights of Quebec. He was his valet, and when MacNeil was wounded with a French bullet it took Calum six long weeks, sucking and cleaning the wound, until one day he sucked the bullet out of the back of his ear. After the War of Independence was over, Calum came back to Barra again but he said he would not stay here any more, and he left for the Island of Cape Breton with his seven sons. And seeing a little corner of Cape Breton, he called it 'The Piper's Cove' because it was so like the little cove he left behind on the Island of Gigha. And, wonderful to say, a great-great-grandson was this year seeing the Coddy and he found out all the information about his forebears that he wanted.) The party landed at Coilleag a' Phrionnsa[3] and, headed by Calum the Piper, the Prince was taken ashore by a man from Eriskay called John MacEachen. Then they opened up a big spread, which consisted of the famous liqueur Drambuie, and the piper played at the spread, and I know the place well.

Afterwards the Prince proceeded to the Laird of Boisdale, a prominent figure in the Islands in those days. And he slept in a thatched house in Eriskay and through the night he was very uncomfortable; the fire was in the middle of the floor and the smoke was cutting his eyes bitterly.[4] Next day he made for Boisdale and he went to see that old, venerable gentleman, the Laird of Boisdale, and told him his mission.

---

[1] *Recte* 'Du Teillay.' [2] i.e. the Gigha in Barra Sound. [3] 'The Prince's Shore.' [4] This house existed until comparatively recently; Miss Goodrich Freer refers to its destruction (1902).

'Oh,' the Laird of Boisdale replied, 'you had better go back and go home – and get ready more men, more ammunition, more money and more food.'

And the Prince replied, 'I have come home, Boisdale.'

'Well,' says Boisdale, 'you will say that before the conflict is over.'

Now the Prince turned to the ship. He was not at all pleased with the decision, to get the sort of encouragement he got from Boisdale. Never mind, they up-anchored and made sail for Moidart, where seven of the chiefs of Lochaber met him, and in a few days the Royal Standard was unfurled at Glenfinnan.

The Fuday crofter was pleased to see the ship coming from Fuday, and he was more pleased to see her going back to the mainland, and in fact he was putting out his chest that he gave three hundred pieces of gold to support the coming of Prince Charles.

The wonderful part of the story is, now, how did that man get so much gold on the little Island of Fuday? And there was a shepherd on the island some sixty years ago who told me this story. And I asked him where did he think the man got the gold?

'Well,' he says, 'I was told, and it was a traditional story of Fuday, that a ship was wrecked on the west side of Fuday. (They had no name for it but he called it the *Long Dhubh* or Black Ship.) And it had a lot of gold in it, and this man Donald MacInnes' forefathers for many generations were on the Island, and were there when the ship was actually wrecked, with the result that they had captured the majority of the gold and divided it and hidden it amongst themselves. And I heard,' he says, 'that portions of it were lost all over the island.'

Now to return to the Prince. His ship crossed the Minch and landed at Arisaig and eight of the chiefs of the clans met the Prince at Moidart and rejoicing over the meeting they danced for the first time the world-famous dance, 'The Eight Men of Moidart.'

[*The late Fr Allan McDonald preserved some information about the MacInneses formerly on Fuday. There were three of them in his time, Aonghus mac Fhionnlaigh 'ic Iain Òig 'ic Fhionnlaigh; Dòmhnall Bàn macPhadraig ic Dhòmhnuill Òig ic Fhionnlaigh; and Ailein Mac Iain 'ic Dhòmhnuill Òig 'ic Fhionnlaigh. They lived then (1897) in Smercleit and North Boisdale, South Uist.*

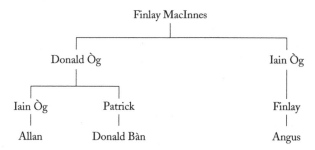

(*exact seniority of the different branches uncertain*).
Fr Allan says that 'McNeil of Barra offered them [*?the last generation*]
three nineteen-years lease of Fuday, aonta tri naoi diag, *but they did not
accept it.' Donald Òg may have been the Donald MacInnes of the '45.
Angus MacInnes was a famous Gaelic storyteller.*]

Coddy's great-grandfather Neil MacNeil and the prisoners of the Napoleonic war

In the Napoleonic wars even from the islands of Uist and Barra
there were prisoners of war in France, and as it so happens, my own
great-grandfather Neil MacNeil was one of them. Also his mate,
John Campbell from South Lochboisdale.[1] Well during their period
in prison they were having a very poor time and they were the only
two Gaelic-speaking men in the whole prison. One day they were
working in the prison yard and their overseer was not far from them,
and strange as it may appear, when they were talking in Gaelic he understood every word of it. And the topic they had this day was complaining about the food. And when I tell you that it consisted of raw
linseed oil and sawdust, you will not be surprised at the complaints.
The prison guard drew near them and said,

'Well, boys,' he says, 'I am hearing you talking Gaelic, and I am
hearing you also complaining about the food. Well, I have no other
alternative but to agree with you that the food is very bad.' And he
turned to my great-grandfather and says, 'What is your name?'

'My name,' he says, 'is Neil MacNeil.'

'And where do you come from?' says the guard.

[1] Possibly Iain Bàn, grandfather of the late Seonaidh Caimbeul, the Gaelic bard, of
South Lochboisdale.

49

'I come from Barra,' he says.

'Well,' says the guard, 'when I heard you were a MacNeil I should have known that you were a Barraman.' And he turned round to Campbell and asked him his name.

'My name is John Campbell.'

'And where do you come from?' he was asked.

'I come from Boisdale, South Uist,' says John.

'Well, that is very strange,' he says, 'my forefathers left Uist in the 1715 rising and we, their descendants, always kept up the Gaelic, and I am very, very pleased to have met you. And I will be very, very pleased to assist you, and if you want out of the prison to-night, that I shall help you to do. And I shall also guide you to the shore,' he says.

'It is your own choice, boys. If you are caught, you will be no more, but if you are not caught you stand a chance. And so I will leave you to decide between you and if it happens you decide to go, I shall leave the gate open for you.'

Now Campbell and MacNeil had a conference and they began to talk.

'Well,' MacNeil says, 'if there is no improvement in the food,' he says, 'there is nothing else for it but to die anyway.'

And Campbell says, 'But if we are caught the guard said we would be no more.'

But MacNeil insisted that it would be better to die in the attempt at escape than to perish there in misery. And so they came to a final decision to see the guard and tell him to leave the gate open and they would clear off through the night.

Well, when the dead of night came they went to the gate and it was easily opened and so they set off on the instructions that they guard gave them to make for the shore. And by the story which I have gathered, it appears that they were not very far from the port and they arrived there pretty early in the morning, and when they went rushing down to the harbour there was an English frigate, with all the sailors aloft among the sails, spreading the canvas ready for the first opportunity to sail. Well, Campbell and MacNeil made a beeline for the ship and did not wait for any orders. They were so pleased to see a ship that they jumped on deck and from the deck up the shrouds, and from the shrouds to the yardarm.

Now the captain's attention was drawn to this, and they would not be allowed to do anything in the ship until he had interviewed them, and they were called before the captain and he began to cross-examine them and he discovered where they were taken prisoner and he discovered also that they had escaped from the prison; they didn't tell why or how. Well, they had a comfortable voyage to England and they were discharged from any more service – the war was about at an end anyway. So Campbell came to South Uist among his people, and MacNeil came to Barra among his people.

*   *   *

Now, after Colonel Roderick MacNeil of Barra came home in 1821, he sent for Neil MacNeil – and I may tell you that he was a blood relation of MacNeil of Barra. And they began to exchange news and MacNeil told how he escaped from the prison, and Colonel Roderick MacNeil explained how he fought his share in the Battle of Waterloo, and after they exchanged news they had a dram together and afterwards MacNeil of Barra gave MacNeil the prisoner a croft on the west side of the island – one of the best crofts on the island – and after, he was there until the estate of Barra fell into the hands of Colonel Gordon, and all the crofters in the township of Greian were removed to the east side of the island, where the worst lands were, to make room for sheep. And there were a lot of emigrants about that time, but MacNeil never went out of the island. He had a family of six daughters and two sons, and one of the daughters was my own grandmother, and Mary was her name. And she is buried under one of the Iona stones at Eoligarry.

MacNeil of Barra and the butler, the gardener and the groom

After the Napoleonic wars were over, General MacNeil, the second-best-looking man on the field of Waterloo, settled at Eoligarry[1] and he employed a groom, a butler and a gardener, as the garden was very much neglected. After the long period of years that it was not looked after it was in very bad condition. So it happened that the butler, the groom and the gardener were staying in a bothy of their own.

One night they planned together it would be a good idea to go into MacNeil's cellar and steal a quantity of whisky. Well, first the

[1] He succeeded in 1822. He was in the second regiment of the Life Guards.

gardener had to be consulted, as the cellar window was looking on to the garden and they must have the key of the garden before they would get near the whisky at all. And he agreed to give the key and to supply a ladder to climb to the cellar. Well, that was settled and it went all right. The butler went up on the ladder and put a considerable quantity of straw on the bottom of the wooden bucket, in case the transferring of the whisky to the bucket made a noise. The groom was supplying the buckets. He took over three buckets and he wanted to take more, but the butler would not allow him, he said the stock was so low.

Now they had the party and they were drinking merrily – in fact it reached a bad stage, almost a disaster. There was an old man there: he was the harper that MacNeil had, and he was a man – although blind – of great bodily strength. The groom went over to him and offered him a dram, and he was not believing him, and he asked him – 'Give me your hand,' he says, and the groom gave him his hand, and, to give you an idea of how strong he was, in grasping the other man by the hand he squeezed blood out of his fingertips. And the groom could not yell, and he could not let on he was suffering pain, in case MacNeil would hear the commotion.

Now they went to bed and slept quietly. Then the groom began to imagine a sort of court – what would happen if it were discovered they had broken into the cellar and raided the whisky. And the gardener got the wind up that he would get the brunt of the fine if it was discovered that he gave the key of the garden so that they could climb up to the cellar. They were for several days planning what to say and what not to say – but fortunately nobody ever knew what actually did happen and nobody ever found out that the whisky was stolen.

However, then they were dry again and still safe after that – the cloud was over and they were beginning to forget about it – and they made up their minds to go to see the gardener about the use of the key into the garden, and the ladder to go and raid the cellar again. But unfortunately he sat down on the stool of repentance and he said, 'Well,' he says, 'I gave you the key already, and take it from me, boys, I will never do it again.'

So they dispersed, and I must not omit to tell you that the gardener was my own grandfather.[1]

[1] i.e. Robert MacLachlan, see p. 13.

The MacLeods of Dunvegan

Dunvegan Castle

The caretaker of the Castle of Dunvegan knew the history of the MacLeods well, of their clan battles and their fights with other clansmen who were attacking them, and she told me stories in this connection too numerous to mention.

Inside the castle are many trophies taken from all parts of the world by the famous chiefs of the MacLeods, including the Fairy Flag and many other interesting articles. There is the famous dungeon, and it is said that one of the MacLeods, Hard-Hearted MacLeod he was called, threw his wife into the dungeon for the reason she told her father in Sleat that MacLeod was going to raid him, and when MacLeod arrived at Sleat, all MacDonald's cattle were safe and the raid was a failure. When he came back in a fury he threw the wife into a dungeon. And any time she screamed for food or drink he used to throw down into the dungeon chunks of salt beef. Latterly the good lady died, and there is a very handsome painting of her in the castle, along with many others.

The castle and the surroundings are the most interesting to visit and one of the special corners about it is on the south-east side, the port-cullis, where you could land if you only knew the way to do it. I was shown that. And for anyone visiting Dunvegan, I would say it is a most interesting sight to see the castle.

[*The dispute between MacLeod and his wife occurred in 1733, see the* Book of Dunvegan, II, 19. *She was Janet MacDonald of Sleat. She objected to her husband's sisters living with them at Dunvegan. According to the editor of the* Book of Dunvegan, *in 1733 she 'posed as a prisoner because she had no money to furnish a house, a statement which may be the basis of the charge that he [MacLeod] put her in a dungeon.' The editor adds that he disbelieves in this story. Janet went to live in Edinburgh in 1733 or 1734, and eventually was reconciled to her husband, on the condition that his sisters were not to be allowed to live at Dunvegan Castle, in 1740. She died in 1741.*]

MacLeod of Dunvegan and the Duke of Argyll

In olden times it was customary for the Highland chiefs to visit each other. This time MacLeod of Dunvegan decided to visit the Duke of Argyll. So one day he got on his horse at Dunvegan and rode across to Kyle of Lochalsh, over the Kyle of Lochalsh he went, all the way to Inveraray. On arriving at Inveraray he discovered that it was a deserted town, and interviewing an old lady at the castle, she told him that everybody in the district, in the whole area, was out at a bull-fight today. And MacLeod said: 'Well,' he says, 'I am very surprised to hear that there is a bull-fight at Inveraray and I shall be glad to hear from you what is the reason.'

'There is a young good-looking fellow,' she said, 'who was accused of some crime – I cannot fully explain – and the magistrates of the area decided that he was to be put to death today by a bull.'

'Ah, yes,' said MacLeod, 'and how far is that away?'

She gave him a description how far and he jumped on his pony and rode to the park.

On arriving, and the Duke of Argyll hearing that the MacLeod of Dunvegan had arrived, he sent for him immediately, and they both had a consultation about the death of the young man. MacLeod demanded that the young man should be brought to him, and after he had seen the young lad he was very much in sympathy with him – he was a Campbell – that such a good-looking fellow would be killed by a bull – and the reason a very frail one.

They began to argue the point very much and then MacLeod said, 'He won't go through with it – I will fight the bull and you can exempt the young lad. His life would be short if he faces that beast, but I shall conquer him.'

Argyll now spoke very strongly against MacLeod's decision, and he said, 'What will happen,' he says, 'to your territory in Skye, if you are killed?'

'Ah,' he says, 'I am not the least afraid of that, Argyll. You let me to the bull and I will do the rest.' And still Argyll was not willing to give in.

* * *

Now he got ready and he jumped into the park, and he wore nothing but the kilt and bare socks, and going straight to the bull, and before the bull could do any performance, MacLeod had the bull by the horns, and his feet well tucked round the bull's neck. Now started the great excitement. The bull went roaring and at a furious pace round the park and the more he roared the firmer MacLeod would hold his horns. Latterly he lay down for a time, snorting and rolling abut, turning the side on which he lay down. And for nearly a solid hour he continued going round and round the park. A second time he fell to the ground again with furious anger, and his bad nature getting the better of him every time. This time he was getting short of breath – but however he made a third attempt. And he did not continue the third round so very stubbornly; he was on the decline all the time, and on the third round he lay down on the ground and could not walk or run another yard, or another inch, I should say. And when MacLeod could see he was stretched out, and gasped for the last – the bull was dead – he let go his horns and came to the ground. Immediately he got to the ground he got his *sgian dubh* and with the *sgian dubh* penetrated the bull to the heart.

Now the bull was dead and MacLeod alive and the people all over the park clapped their hands. Now MacLeod took his breath and looked at the bull stretched out, and his next thought was, as proof of what he did, he would take the bull by the horn and pull the horn by the root out of his head. And this he did.

And then he went over to talk to Argyll who congratulated him, and so did everyone in the arena. And he wanted the boy in his presence, and shook hands with him very affectionately, and the boy thanked him very kindly. And MacLeod asked him, 'Will you come with me to the Isle of Skye?' And the reply was 'Yes, sir,' he says, 'I will go with you anywhere.' And after a pause it was finally decided that he would take the horn with him to the Isle of Skye, and the boy to the Isle of Skye.

To the day I heard the story, in the Island of Skye, descendants of that boy were at Dunvegan, and MacLeod gave an order to convert the horn into a drinking cup, so that whenever any MacLeod was proclaimed a chief he had to drink out of the drinking cup. I have seen the horn myself in the Castle of Dunvegan. I collected this story from the shoemaker not far from the castle in 1911.

[*This would be the famous drinking horn of Rory Mór MacLeod of Dunvegan who died in 1626. At that time the chiefs of Clan Campbell had not yet been raised to a Dukedom.*]

The Laird of Boisdale

The Laird of Boisdale and the bag of meal

After the failure of the potato crop in 1846 times were very thin on all the Outer Islands, but especially on Barra. And the harvest of this particular year which I am going to talk about was very dark. The only place where crops were successful was in South Boisdale. Now there was a man in Brevig called Donald MacDonald, crofter, and early in April he said to his wife, 'Well, Mary,' he says, 'I don't think that we shall have as much corn as will do us until we reap the next harvest, and so my suggestion is that I should go across to Boisdale to see the Laird of Boisdale.'

'Well, if you think that is the right thing to do,' she said, 'Donald, I don't object to it. You get up tomorrow morning early, and that is the bag, and that is the last guinea in the house. And I hope you will take every care of yourself and come home with the bag full.' So he made for Eoligarry, from where the ferry was leaving.

Later on, it was decided that the ferry was not going across that day. Unfortunately, now, Donald came across one of the loafers[1] which are often met with at a public house. He invited Donald in to have dram and he had a sixpence – this man who invited him inside. Now they had a mutchkin, and so Donald partook of the dram and so did the other fellow, and then when it began to warm him up he felt inclined to call a drink himself. And this he did, and he broke the guinea that was going to buy the bag of grain. And every drink that Donald was calling they were drinking, and the toast was 'Here's to your very good health, Donald, and I don't think that a man of your heart will ever want.'

And so the boys went on, and Donald not being accustomed to take whisky but very rarely, it made him very free with his money. This continued until the evening, and they went home to their dinner, and the wife of the Eoligarry lad was very much annoyed

[1] *'Sgimileir'* in Gaelic.

at taking the poor man who came to go to the mill for his wife into the public house, and starting him to drink. 'Never you mind,' she says, 'the meal will be all right.' And the dinner was a great favourite of your own – fresh spotted flounders and potatoes. So they both had a good feed and after a rest they went back to the bar again – and they came home late at night and the guinea was spent, unfortunately.

Next morning the boat was going to leave for Boisdale, and Donald said: 'I can't go,' he says – 'and it is very difficult for me to go back home after spending the guinea and going home with no meal.'

'Well,' MacNeil said, 'you will go over to the Laird of Boisdale, Donald, and tell him that you met me here, and that we spent a happy day together; and ask him that I told him to give you a bag of meal and to send one to myself. And as sure as you are sitting there,' he says, 'The Laird of Boisdale will do that at my command.'

* * *

Now they were aboard the boat crossing the Sound and they had a comfortable trip across the Barra Sound, and when they arrived in Boisdale it was their dinner hour. Well, Donald sent word to the grieve and the grieve came to see him.

'I am Donald MacDonald,' he says, 'and I have a message to the Laird of Boisdale.'

'Well, just wait for a few minutes,' he says, 'and I will go over and give the Laird of Boisdale the message. And you walk round to the front door and he will come out without much delay.'

Now there is Donald standing at the front door, and Boisdale comes out, and he says, 'Did you come from Barra?'

And Donald says, 'I did, and I have a message for you from John MacNeil of Kilbar.'

'Oh, yes,' he says, 'and what is the message?'

'This is it,' says Donald. 'I came from Brevig yesterday morning and the ferryman didn't come across, and I spent the day with MacNeil in the pub at Kilbar – and I spent also the guinea I had for getting a bag of meal. And MacNeil got the lion's share of it. And he told me,' he says, 'to come over here and see the Laird of Boisdale and tell him to give me a bag of meal and send him one also.'

'Oh well,' says Boisdale, 'that is very like a thing MacNeil would do. You go to the grieve and tell him to give you two bags of meal – and that is all you need do. You take one for yourself – the other one give to MacNeil when you arrive on the other side.'

I need not describe to you how pleased poor Donald was and how gratefully he thanked the Laird of Boisdale for his enormous kindness to him and wished him every blessing of the season and a bumper of a crop the next year. Those were his prayers on his departure.

Now everything was said, and all the meal was sent aboard and the boat sailed for Eoligarry. And on their arrival who was awaiting them but MacNeil, and he hailed out, 'How did you get on, Donald?'

Donald replied boldly, 'Very well, 'he says, 'and I got a bag for you, too.'

'Didn't I tell you,' says MacNeil, 'that a man of your big heart will never want?'

Now Donald got his pony ready and put the sack on his back, and came home to Brevig, and his wife was glad to see him – but she was never any the wiser that Donald spent the guinea.

[*The date ascribed to this story is impossible. The last Laird of Boisdale, Hugh, who succeeded in* 1818, *was an absentee, and the estate was sold to Colonel Gordon around* 1840. *The incident may have happened in the days of Colin MacDonald of Boisdale who was Laird from* 1768 *to* 1799. *See Fraser Mackintosh,* Antiquarian Notes, *pp.* 323, 324.]

Stories of olden times

The weaver of the castle

The Weaver was banished from Barra to the Stack islands. He took with him a small boat and an ancient *cas chrom* and other implements for cultivating the island. The first thing he did was to go across to Eriskay and get hold of a fair pony, or *láir bhán* as it was called. He then started to build the castle, with stones collected from the shore at the foot of the cliff. He then started to carry the stones by every means he could use, including his back, up the cliff. And to this day you can see where he tipped the pack of stones with the white pony. It took him a long time carrying the stones and building the castle, living on fishing and fowling and what he could produce from the island. And when that failed, he went ashore and helped himself – raiding was common enough in those days.

Now it came to the end of the tether – the castle was finished and the Weaver decided to take a wife to himself. This was in the month of July, when it was the custom in those days of the crofters in South Uist to go to the hills, taking with them the various kinds of cattle, from the milking cows to the small calves. During this period most of the butter and cheese was made for the winter use. The wives and daughters who acted as dairymaids followed them to the shielings, and when the Weaver had a conference to himself where to look for a wife, he decided to go to the nearest shieling to him – which was Loch Eynort, South Uist. At sunrise, maybe before, the Weaver left the Stack Islands, and was at the shieling among a gay crowd of good-looking ladies, with their wooden buckets, getting ready to start the milking. The Weaver made a quick decision for a choice and without much debating he flung this young lady on his shoulders and made a bee-line for the boat, which carried them both safely to the castle.

We now leave the Weaver in the castle with his wife, whom he trained to be equal to himself in raiding. She used to go fishing and raiding with him. In those days it was all cable and hawsers that were used instead of chains, and the Weaver and his wife used to cut the hawsers and let the ships drift to the shore, and themselves getting the benefit of the wreckage. This was the routine until the boys [were born and] grew up one by one and helped him with the piracy, which was now getting to a dangerous point.

Now an order was passed to destroy the Weaver, or apprehend him. One day the Weaver was fishing with his three sons from the castle. The mother was at home with her youngest boy. They both could see the boat fishing. Later on in the day she saw a sailing ship off the Island of Gigha.[1] The ship was a cutter, manned by twelve oarsmen. At first she thought they were making for the castle and she got ready to go to the top of the cliff where there was always a cairn of stones ready for anybody attempting to climb the cliff.

Unfortunately they made for the boat that was fishing. As soon as the Weaver picked up the cutter he made a bee-line for Eriskay. The Weaver and the boys pulled well and hard to make for a point to land on Eriskay. If they could manage to land there he could hide in safety. He did manage to land, but the cutter was there immediately. The commander landed without delay and with his sword slew the Weaver and his three sons, and ordered that the blood of the Weaver should remain on the sword to dry, as proof that that was the sword with which he killed the Weaver. To this day that landing-place is called 'The Cove of Disaster' (*Sloc na Creiche*).

The news circulated from the Island that the Weaver and his three sons were killed on the Island of Eriskay by the commander of the ship. Little did the mother know of the sad event at the time, though she did see the cutter passing out. The next stage was the funeral of the Weaver and the sons. When all was completed, her father went to the castle to take his daughter home and the little grandson. That day and since, the castle has not been occupied. It stands on an island, commands a magnificent view of the Minch, Barra, South Uist and Eriskay.

The life story of the little weaver

His grandfather, with whom the boy was living, dearly loved the boy, and his activity at an early age much interested the old man. At

[1] *i.e.* the Gigha in Barra Sound

the early age of twelve he used to be wonderful in attending with his grandfather on the croft, about the sheep, cattle and horses. When he grew to the age of fourteen he often wondered why his mother used to cry every day. He became so interested that he insisted on his mother telling him the reason why. His poor mother told him the story about the sad end of his father and three brothers. Pausing for a bit, and taking a long and deep breath, he said, 'I am going to sea, and I shall never stop until I meet the man who killed my father and my three brothers.' His mother at this stage broke down worse than ever.

However, the time was moving along and John was daily making up his mind to go. One day he decided to have a meeting with his mother and grandfather, and told them he was going. This was a very sad parting.

In those days there were no conveyances. John had to walk all the way from mid-Uist to Lochmaddy, over two fords. The only connection to the southern isles was by a ferryboat from Lochmaddy to Dunvegan. However, he assisted the boatman, and in return the ferryman was very kind to him and gave him his food and passage charge free. The passage across to Dunvegan was quite good and, landing there, he was much interested in the number of trees he saw, whereas there were none in the land he left behind him.

John stayed with a crofter, working on the croft, and the crofter assisting him a lot as to the right road to Kyle. On his departure he charged him nothing and gave him a little money of the very small amount he had himself.

John was moving across Skye. Until one day he landed in Kyle and got the ferry across to the mainland. John stayed a week in Kyle, working with an old carpenter who was once upon a time a ship's carpenter. John MacNeil overstayed his time, listening to the carpenter's stories. John was keen to find out from the carpenter where was Greenock, as it was at Greenock he intended to get a ship. The carpenter told John he did sail several times from Greenock, and encouraged him by saying he would have no difficulty in getting a ship from there. From the time John left Dunvegan till he arrived at Greenock he covered a full year, walking and working, just as he found convenient. However, the day he arrived at Greenock he was thrilled with the sight of the ships, with their high masts, yards, sails, et cetera. He walked straight down to the harbour and nobody even spoke to him. Having saved a few shillings, he was able to take a night's lodging. Next day he got up early and went down to the harbour. He was not long there when the captain came on deck. Having seen John MacNeil there yesterday, he hailed in English, 'Do you want to go to sea?'

John could not speak but very little English, and did not reply. Immediately then the captain spoke in Gaelic and John replied immediately, 'Yes, I want to go to sea.'

'Come on board,' the captain replied.

The captain was a very good Gaelic speaker, and into the bargain a native of Arran. He asked John what part did he come from. He replied that his father came from the Island of Barra but that he was born on the Stack Islands. The captain then asked his name. He replied: 'My name is John MacNeil.' Then both MacNeils shook hands and ever since that they were the best of friends.

This voyage was to be from Scotland to Vancouver Island, round Cape Horn – quite a long voyage in the sailing-ships of those days. The ship was taken to the Tail of the Bank and the cargo was sugar. She anchored on the Tail of the Bank and the boys got ready to go aloft to bend the sail and get ready for sailing, whenever everything was ready.

John was the youngest on board. Never mind, the material was in him and he was not long picking up. The captain patted him on the shoulder when he came down on deck and told him to take care of himself. 'One day you shall be master of a ship yourself.'

At this stage it took them a few days to get ready. When all was in order, with wind and weather favourable, the order came to stand by and heave the anchor. This was, and used to be, a great time in the old sailing ship, heaving the anchor: no definite time where or when they would hear again the order 'Let go the anchor!'

★ ★ ★

As time was getting on, the captain was getting fonder of MacNeil. He used to watch him with pride and admiration – how very handsomely he would walk the plank, or in other words, walk the decks. Watching him daily he could see how handsomely and skillfully he could run up the shrouds. Before MacNeil was very long in the ship, the captain used to send him up the rigging right on to the royals, where he would stand on the yard and put his hand on the top mast and wave to them all on deck.

The whole voyage out Captain and John MacNeil became great friends, so much that latterly the captain took in hand to teach him the alphabet. Not a long period passed before John could read and write. The captain's ambition was to teach John all the schooling he could. As for the seamanship part of it, John had it all on his finger-ends already. When the ship returned on this voyage the captain sent him to school and left him behind the next one. On the ship's return John sat the exam and passed straight out. He was pleased to be sailing the next voyage as second mate along with his good friend Captain MacNeil. John very much enjoyed his first job as an officer, and I could vouch with safety that nothing was left undone. A few voyages after that John sat another examination and got First Officer's. One day the captain said: 'Now be getting ready, John. I shall soon be retiring, and you shall be taking over full command. But I am not going to take this step until we both agree that all is in order.' The day, alas, arrived and Captain John took over command of the ship and the faithful friends went out and sailed together for the last time. After the voyage was completed the captain and John parted. The good captain went on his knees and kissed John's hands and wished every good wish in his new career – that is how two most dutiful friends and good sailors parted.

*   *   *

Now that is the little Weaver in full command of a full-rigged ship after completing his time. Captain MacNeil was now sailing in the same Company for several years but did not seem to meet or hear any news of the man who killed his father and three brothers. Coming home on this voyage he wondered a lot if ever he was going to meet him.

On arriving in London he went to a club. There he found a lot of old veterans telling stories and drinking whisky. It suddenly grasped on him that among this crowd he might meet the man who he was really looking for – the man who so brutally killed his father and three brothers.

Suddenly one veteran stood up and told the story of how he destroyed a very destructive raider and his three sons on one of the western islands off the west coast of Scotland. One could imagine the thrill Captain John got when the sad story was renewed to him.

The veteran got a lot of cheering. Captain MacNeil went over to him and specially thanked him for his great bravery. In turn the old man cordially thanked him, and invited him to tea next evening at 5 p.m. and told him that the bloodstains were still on the sword with which he slew the raider and his three sons. John left the club and returned to his ship and trying to decide how he would destroy the commander without the use of arms. So he finally decided that with one good blow of his fist he would do the job.

The appointed time arrived. Captain John arrived. The veteran answered the bell and both entered the house. Plenty of food and drink was prepared. Captain MacNeil said he would not eat or drink until he would see the sword with which he destroyed the dangerous man in the Western Isle of Scotland. The old man immediately invited him to his parlour. He opened the cupboard and took out the sword, and bloodstains were still on it as he described. Captain John gave him time to return the sword, at the same time he decided not to kill him with the sword with which his father and three brothers had been slain. As the old man was stretching himself, Captain John struck him right in the ear and the old man never breathed another.

This put an end to Captain John's ambition. Now there was nothing for it but to face the return journey home to Uist. Off he set and left the ship and cargo there. The return journey did not take him so long and one day he found himself landing in Lochmaddy. It happened to be a fine summer day when he arrived home. Sitting outside the door was an old lady who he took to be his mother – and he was right. When she saw the boy dressed in blue approaching the house she rose to meet him and when she came within speaking range she asked him: 'Are you a sailor? Or did you ever meet my boy?' At this stage John jumped to and kissed his mother, who was speechless for a time.

Captain John stayed with his mother in Uist for over two years – until times calmed down; then sailed out of Liverpool, where his descendants still flourish.

[*There is a long and circumstantial version of this story in the papers of the late Fr Allan McDonald of Eriskay. It was probably taken down in 1893. The name of the reciter is not given and the story is told in Fr Allan's own words.*

In this version the Weaver is said to have come from the mainland,

and to have acquired his nickname from having married a weaveress from Kildonan in South Uist. It describes how he took the white mare to Stack to carry the stones to build the 'castle' and how the mare fell dead from exhaustion with the last load, the contents of the panniers remaining as two cairns, as can still be seen. In this version the Weaver had three sons, and the only stranger who ever visited the castle was the midwife who was brought to deliver them. In consequence of the Weaver's depredations the king sent a boat to capture him. He and his two eldest sons were caught by a ruse and put to death.

When the youngest son, John, grew up, he resolved to avenge his father. He made his way to Dunvegan via Lochmaddy, and there learnt that the man who had killed his father was captain of a ship sailing between Dunvegan and Tobermory. He waited until this ship entered Dunvegan harbour, boarded it, found the captain in possession of the bloodstained sword with which the deed had been done, and killed him.

At this point the story takes a fantastic turn and becomes mixed up with events which belong purely to the folklore of the old Gaelic stories. John took a job with an inn-keeper, was told to guard the garden against deer, aimed his gun one night at a deer which turned into a woman, who told him she had been be-spelled by the inn-keeper. An assignation was made, but three times frustrated by a sleeping-apple which John was persuaded to eat when smitten by thirst. At the last encounter the lady left John a ring and a knotted handkerchief, and wrote with her fingernail on the stock of his gun that her name was on the ring and that whenever he unloosed a knot in the handkerchief he could get a wish.

After waking, John set out in search of her, and eventually learnt she was in the Kingdom of the Great World (Rìoghachd an Domhain Mhóir *of folk-tales). He arranged to get carried there in an oxskin by a griffin. He escaped from the griffin's net by unloosing a knot in the handkerchief, and, learning there was to be a celebration at the palace that night, attended it, was recognised by the lady* (who was the King's daughter) *through the ring, married her and lived splendidly ever after.*

There is an inferior version of this second part of the story printed in More West Highland Tales, *p.* 394, *the annotator of which has been handicapped by ignorance of Uist traditions, dialect and topography. This was taken down from Patrick Smith of South Boisdale in* 1859. *It is difficult, indeed, to believe that Patrick Smith, who was a famous traditor, did not really know this story in full.*

As regards the main part of the tale, it is very probably founded on fact.

The Stack Island, which I visited with the Rev. John MacCormick in 1951, is shaped like the figure 8. The isthmus which joins the two halves is very narrow and on the south side is faced by a cliff up which there is only one path, easily defended. There is good grazing on the island. At the top of this path are the two small cairns said in the story to be the last load of the white mare. On top of the Stack, commanding a magnificent view north and south up and down the Minch, and over Barra Sound out to the Atlantic, are the ruins of a small stronghold and of a small house wall. The 'castle' is made of very strongly mortared stone and its appearance suggests that it was destroyed by gunpowder.

Ships used to anchor in Barra Sound in the lee of the small islands of Hellisey and Gigha, and it was these of which the cables were cut by the Weaver and his sons rowing out under cover of darkness. When the ships were driven ashore, they would be plundered.

That this kind of thing did occur in the late sixteenth and early seventeenth centuries is proved by various references in the register of the Privy Council of Scotland. In 1611 the Barramen were in trouble over the pirating of a ship, laden with Bordeaux wines, belonging to one Abell Dynes. In 1636 there was similar trouble over an English ship, the Susanna, *which had been bound for Limerick with a cargo of wines, fruit, coin, etc. Gales had driven her out of her course into the waters of an island (not named). She had lost mast and small boat, so she signalled to the islanders who came out to her armed to the teeth. They agreed to tow the* Susanna *into safe harbour in consideration of a butt of 'seek' (?sack) and a barrel of raisins; but it was alleged that, after they had cut the anchor cable and brought her into harbour, Clanranald and about three hundred others took casks and barrels down to the shore and daily drew of the cargo of wine, and took all else besides, robbing the crew even of their clothes, and that then they made some ship's lad sign a document to say that he was owner of the cargo which he thereby sold at a certain price (which amount he did not in any case receive), and under threat of handing the whole company over to the 'savages that dwells in the mayne,' the captain was forced to sell for £8 the barque worth £150 sterling.*

So that the activities of the Weaver and his sons, his death and the revenge exacted by his youngest son, are perfectly probable. The folkloric elements are likely to have been tacked on to the story of young John by later storytellers in order to entertain their audiences. They are found in quite a number of tales.]

The emigrant ship Admiral *and the Barra evictions*

The clearances of South Uist and Barra were due principally to the failure of the potato crop of 1846. After that a lot of hardships came over the islands; and it came in one night, the blight, and covered the whole of the potato crop. And I remember clearly seeing a man who made up his mind to leave the Island of Barra. His name was Neil MacNeil. The smell that was off that blight was enough to choke you, and so he decided to pack his bag and went to Castlebay to get some conveyance – a boat or ship – to take him to the mainland. And that he did. And that was his description of what happened the night of the blight. It was about the 14th August. Well, he left, and it was fourteen years before he returned to Barra, and several who went with him never did return.

(During Neil MacNeil's long periods of absence from Barra he was working in many queer places all over Scotland, including the making of the Caledonian Canal. His wages were a shilling a day. I said to Neil, 'Well, Neil, the wages were very poor.' And Neil said, 'No; the stone of meal,' he says, 'cost only a shilling, and the half mutchkin of whisky cost another shilling, and then you had five shillings to yourself.')

Now, that left a very bad mark on the population, and a mark of poverty, and Ground Officers came and went round the crofters and told them that they would get so much meal, oatmeal, if they gave their names in to go to Canada. A great many were pressed to go – but they wanted their names before they gave them their meal, on the condition that if your name was there to go, to Canada you must go.

Latterly, the unfortunate day came round and the ship came and anchored in Lochboisdale, and the orders came that all those people who got the meal for going to Canada had to get ready to embark and clear out – a disastrous story. In those days many a sad story would be told when the people were clearing out. I remember myself two old ladies, two sisters, that signed with their father to go, and they took to the hills and they were hiding up in Ben Cleat and coming down at night to Vaslan to friends' houses who used to give them something to eat. And this was carried on until the coast was clear, and their old father went. And the unfortunate part of the story, you see, he died on the way out. His name was John

MacDougall, known in Barra as Iain Muilleir, and he was buried at sea.

All the people from Barra were ferried over in boats from Barra to South Uist, and I remember clearly a man, Archibald MacLean, son of Neil MacLean, who was on the way. And he did not manage to be ready when the ferry-boat went from Barra, but he chartered another boat so that he would be there before the ship left Lochboisdale. They arrived at the other side; going up a hill at Polacharra they could see a ship, the *Admiral*, sailing round and out past the mouth of Loch Boisdale with three hundred emigrants on board, off to America – so that was the voyage to America completed for him! Archie I knew well. He was a blood relation to myself – his mother was a MacPhee – and he lived until he was a hundred and two on the Island of Barra.

* * *

Now I am going to describe the scenes that took place while the emigrants were put aboard and the brutality shown to human beings was beyond description. I knew a man by name Farquhar MacRae, who was on board the *Admiral* and witness to all that was happening. He explained to me very fully, although I was young at the time, all that happened when the friends and relatives came to see them off.

I am going to mention one man in particular, Donald Johnston, South Lochboisdale. The Ground Officers volunteered to tie him down and he stood up there like a hero and he said very distinctly that all the Ground Officers from the Butt of Lewis to Barra Head would not tie him or handcuff him. So they had to abandon the scheme. Well, the priest was there. He went over and tamed Johnston and calmed him down and and said to him that he was very confident that no Ground Officer could compel him to go. 'But,' said he to him, 'you will go of your own accord and I will give you my blessing.' And that was done and he went.

Now the worst confusion of all started aboard the *Admiral*, and MacRae said it was a sad and horrible sight to see the parting of those that were going from those that were going to stay behind. The captain was a burly big fellow and he gave orders to erect a sort of fence amidships, with those that were staying on this side of the fence, and those that were going out on the other side of the fence.

And after he got this done he gave an order to the sailors, 'Come on, boys, now lash them out.' And the sailors got rope and lashed the poor men and women into their boats back home, after parting with their nearest and dearest friends.

Now they were aboard, and they sailed away and it took them six months, I think, to get across from Lochboisdale to Quebec – or the St Lawrence somewhere, anyway. After that there was nothing for them but trees and poverty – no food and no provisions made for them, and those wretched fellows had to back it for life.

Going through the hardships of the emigrants who went across the Atlantic in the *Admiral*, not many people would credit them today.

★   ★   ★

And now the fourth generation of them came to Barra last August and they were supposed to meet the Coddy, who gave them full information where their forefathers lived on Barra. And one of the places that they had in mind was the Cooper's Beach, and another had in mind Gighay. The Cooper was one of the .principal men on the island in his day. He was a cooper in the glass factory in North Bay. And the other man – his forefather left the Island of Gighay, and I mentioned his name in another story – he was the pilot who did bring the Ladoutille and the Prince and his followers and landed them on the beach at Eriskay.

A great-grandson of the Cooper visited Barra this year, 15th August, and his forefather landed in Cape Breton, and some of his descendants are on that farm still. But for those who landed it was with a great struggle that they managed to have an existence, though today they are comfortably off and the majority of the Cooper's descendants are well off in many parts of America. And so many of the descendants are well off in many parts of America.

Also many of the descendants of Malcolm MacNeil the Piper are in important businesses all over the States and Cape Breton. There is a bay in Cape Breton, called Piper's Cove, and it resembles very much the place in Barra from which the Piper emigrated.

★   ★   ★

One year the Piper and his sons were going to fish and they ordered a smack from the Island of Arran to come to supply them with salt. Calum went aboard with his seven sons to discharge the cargo. Now they set up planks and made stages to unload the boat, and they made the stages in a position that the one man could hand the other the barrel of salt, and so hand over hand until it was piled on the shore.

The Arran man was so horrified at seeing the activity of the men discharging the boat that he did not tell the boys how much the salt cost. He said, 'Goodbye, boys, we will call to see you again next year.'

[*Coddy's account of the atrocious evictions of 1851 is strikingly confirmed by the evidence given to the Crofters Commission by John MacKay, crofter, of Kilphedir, South Uist, on 28th May 1883. MacKay was seventy-five years old at the time and had seen these incidents with his own eyes. He submitted a paper on the grievances of the Kilphedir crofters to the Commission. Later in the course of his cross-examination the following exchanges took place:*

Chairman (Lord Napier):*You make a very serious charge in this paper which requires a little explanation. You say, 'Others were driven and compelled to emigrate to America, some of whom had been tied before our eyes, other hiding themselves in cave and crevices, for fear of being caught by authorised officers.' Will you explain these words?*

MacKay: *I heard and saw portions of it.*

Lord Napier: *Will you relate what you heard and saw?*

MacKay: *I saw a policeman chasing a man down the* machair *towards Askernish, with a view to catch him, in order to send him on board an emigrant ship lying in Loch Boisdale. I saw a man who lay down on his face and nose on a little island, hiding himself from the policeman, and the policeman getting a dog to search for this missing man in order to get him aboard the emigrant ship.*

Lord Napier: *What was the name of the man?*

MacKay: *Lachlan MacDonald.*

Lord Napier: *What was the name of the previous person you referred to?*

MacKay: *Donald Smith.*

Lord Napier: *Did the dog find this unfortunate youth?*

MacKay: *The dog did not discover him, but the man was afterwards discovered all the same. He had got into the trench of a lazy bed.*

Lord Napier: *What was done with him?*

MacKay: *He was taken off.*

Lord Napier: *And really sent off like an animal that was going to the southern markets?*

MacKay: *Just the same way.*

Lord Napier: *Did you hear that the same thing was done to others, although you did not see it?*

MacKay: *A man called Angus Johnston, whose wife gave birth to three children, and another child was dead before, he was seized and tied upon the pier of Lochboisdale; and it was by means of giving him a kick that he was put into the boat and knocked down. The old priest interfered, and said, 'What are you doing to this man? Let him alone. It is against the law.' The four children were dead in the house when he was caught and tied, and knocked down by a kick, and put on board.*

Lord Napier: *Speaking generally, are you able to say from hearsay that you have no doubt in your own mind there were many other hardships and cruelties committed in the course of these evictions?*

MacKay: *Yes, no doubt. . . .*

In subsequent evidence MacKay said he had known Angus Johnston and his wife. He himself, when in charge of a gang of men working on the roads, had been asked to bring people out of their homes to be sent away on an emigrant ship, but had refused.

Michael Buchanan, the chief Barra witness who appeared before the Commission, said that during the evictions that had taken place about thirty-five or thirty-six years earlier he had seen with his own eyes 'the roof of a house actually falling down upon the fire, and the smoke issuing."

It need not be thought, therefore, that there is any exaggeration in Coddy's account.]

Inns and ferries, and MacPhee the ferryman's daughter

In the reign of Charles I there was an Act passed in Edinburgh that there would be an inn on both sides of every ferry in the whole of Scotland. And this was carried out and it was proving very satisfactory for travellers, although I believe that in some instances it was abused. At the time I am talking about, there was one in South Uist and one in Barra, and the one in South Uist is still in use though the one in Barra was dismantled many years ago.

Now at this time letters were conveyed to Barra by a ferryman from this side called Archibald[1] MacPhee. He had eleven daughters

[1] Alasdair according to the Baptismal Register. See Coddy's family tree.

and three sons, and one of the daughters happened to be my own grandmother, and her name was Flora. The ferryman had only four letters coming to Barra every month – one for MacNeil, one for the minister, one for the doctor and one for the priest, and the postman was illiterate. Well, when they arrived at Eoligarry the postmaster said to the postman, 'Now, Peter, that is for MacNeil of Barra' – and he had a pocket especially for him – 'and that is one for the doctor in that pocket, one in this pocket for the minister, and one for the priest in that one.'

On Peter's arrival at home he would take the letters out of his pocket and put them below a different coloured bowl on the dresser, and at the same time repeating to himself, 'That is the letter for MacNeil of Barra,' and putting his hand in the other pocket, 'That is the letter for the doctor,' and in the third pocket, he says, 'That is for the minister,' and the last one, 'That's for the priest.' and they were to lay below the bowls that night and the next morning Peter was gong to make the delivery. As soon as ever he got up in the morning and got something to eat, he would go to the bowls and say: 'That is MacNeil of Barra's letter . . .' and so on, to the doctor, the minister and the priest. And then he would start his famous delivery. One can understand easily that owing to the fact that letters came from Dunvegan to Lochmaddy once a month, correspondence was rare and not heavy.

★　★　★

On one occasion there was a married woman in Skye in a place called Tallisker, where the well-known whisky is made. She had three of a family and she was very keen to come to Barra to baptize the children. Well, one day she was in the house, and her husband was working outside and the boy was seriously ill. So she vowed that she would not pass another winter in the Island of Skye, if a priest did not come, without coming to Barra to baptize the children. So she put it to her husband and they both agreed that it was a good proposal. As soon as the first of May came, she left Tallisker and walked to Dunvegan – a considerable distance. Next day she was fortunate to catch the ferry to Lochmaddy, and she walked from Lochmaddy to Polacharra, South Uist, something over sixty miles, got the ferry at Polacharra, came across the Sound and landed at Eoligarry, and walked from there to Craigston, being the only church in the island at the time. And when they arrived, the priest was at home and she told the story to the priest which I am after telling you just now, and one of the daughters he sent to the well with a little bowl in her hand, to take home the water with which he baptized her, and also the two other children, a boy and a girl. So that was very dutiful of her – and perhaps it will surprise you to know that that was my own grandmother.

[*It was the second of the Statutes of Iona, imposed upon the Isles in 1609, which ordained the establishment of inns at convenient places. This, of course, was in the reign of King James VI and I.*

The journey of Coddy's grandmother from Skye to Barra took place in 1848; the Register shows that the baptisms described took place on 21st June of that year. See Introduction.]

The shebeeners of Kintail

In olden days the shebeeners of Kintail were famous throughout Scotland. The following is a story of Donald and Mary MacRae, both of Kintail. When the time matured, Donald went to the hills and prepared his shieling which was always in the same place and near a beautiful stream of the finest water that flowed to the waters of Kintail. At the appointed time Donald takes his grain up and stores it in the shieling, and then, when the peats were ready, and all was ready in the shieling, Donald took up the *poit dhubh* and started to distil the whisky.

He had also with him a grey mare with two creels on her back, and inside the creels were to be two casks and the two casks must go to Mary down to the *clachan* in Kintail. And Mary was always ready to stand by and give Donald a hand unloading, and then Donald took off the creel, and Mary supporting it on the other side. When the creel was taken down there was a quantity taken out of it, and both Mary and Donald drank a bowlful so as to prove that it was up to standard and strength.

This continued for a long time and it was getting towards the end of the season and Donald was getting short of grain, with the result that he had to put more water into it, thus reducing the strength of the whisky. When Donald found himself in this predicament he said to himself, 'Well, Mary will be none the wiser – I will fill the casks by putting more water in than usual.' And he got the load ready on the back of the horse. And it was the last of the year if he did not take up fresh grain. Well, he was going to consult Mary about supplies for the future.

Arriving at Mary's house they took down the sack as usual, opened the cask and sat at the bowls, and Mary did not enjoy the drink very long before she said to Donald, 'Donald,' she says, 'I am not going to buy from you today at all.'

Donald knew he had committed a crime but he said, 'I am surprised to hear that.'

And Mary replied, 'On no account will I buy it.'

'And will you tell me why, Mary?'

'Well,' she says, 'you put too much water in it.'

'Oh, that cannot be, Mary – I just put the usual amount in it.'

And, 'You did nothing of the kind, Donald.'

And, 'How can you tell me, Mary?'

'Oh, yes,' she says, 'when I used to take your dram, Donald, out of the bowl from the cask, it put a certain warm glow in the boundaries of my navel – and there is no kick in it today, Donald.'

Well the argument was still continued and Donald had to make his confession and tell Mary that the reason he put a drop of extra water in it was that he fell short of grain

'Oh, well, Donald,' she says, 'that is all right, but we shall reduce it out of the price.'

This was done and the price was reduced and Donald asked Mary, 'Are you going to take any more this season?'

'Well, yes,' she says, 'Donald, but the demand is now getting poor for it and the season is more or less finished. You will take another supply down but be very careful you don't put in the water to the extent you did the last time. You and I built a good reputation and I have a lot of good customers over your quality and it would be bad for us to lose our customers through supplying bad stuff.'

So Donald promised that he would not commit the same crime again and he went back with a good load of grain and he came back with the casks well filled, and over and above well filled, and over the proof of anything! Now he said nothing to Mary, but took off the sack. They opened the cask and filled the bowls. They began to take it – and Mary, feeling the quality registering properly in her bowels, she laughed merrily and said to Donald, 'Well Donald, my dear

Friend,' she says, 'you looked after the blending very well this time.'

That was the end of the season and when the time matured next year, they began selling all the whisky and supplying the usual superior quality.

The drowning in Barra sound

Once upon a time there lived in the village of B. four fishermen, they were the boat's crew, very capable and excellent fishermen.

One winter they began to trawl a lot of saithe up at Bolnabodach and in very good time they filled the boat and then got themselves in readiness to go over to South Uist and sell it. Now they sent a message to my father to go with them, and he got ready to go, and on the way from home to the boat, you see, he felt something telling him to come back and not to go at all – and he didn't go. And they left Barra with a full cargo of saithe and landed at Polacharra, South Uist, and got good sale for their fish.

After they had sold their fish, they made for Barra. It was a beautiful moonlight night, but half-way coming across there came a shower of snow, and it appears that the wind blew up and the squall that came filled the boat. She didn't actually capsize but it was full of water and the bailer they had could not be found, with the result that they spent a very severe and cold night.

Now in the morning early, there was a little boy on the Island of Fuday, and he said to the shepherd who was there that he was seeing a boat that appeared to be under the water coming down the Sound. This shepherd came out and looked at the boat and knew it, and he understood that they were very much in difficulty. The tide was ebbing, going out to the Atlantic, and there was every danger that the boat would miss the island and go out to the Atlantic.

The shepherd's boat was high and dry and he ran down to a point that he was almost sure the boat would touch, and he got a rope and he put a stone on the end of it and threw it into the boat, and with that assistance he managed to pull the boat to the shore and with difficulty he took out the three men, and the fourth man was missing. He took the bodies ashore and he put them just lying on the beach but in a place where the tide would not get near them, with the assistance of his wife and the little boy. Then they put out the boat, and the shepherd when he came to the mainland told all the people about it. It happened to be on a Sunday, and the shepherd went to the Church at Craigston and told the priest to pray for the repose of the souls. Well, that was done, and very quickly there was a boat with many people in attendance at Fuday.

And I have often heard my mother talking about the gloom, and the sad gloom, that covered the whole island.

Now the day of the funeral came and it was the biggest funeral that was ever seen at Barra, three coffins going down the road. I clearly remember seeing my mother weeping and wailing as the crowd went past, and during my own career I happened to be ferrying a lot of people in the 1914 war across there, but she never ceased to advise me to keep it hot in my ears – the disaster that occurred – and always be careful in my boat. I am proud to say that I did follow her advice, with the result that I am still alive and did not meet such a fate.

Everyone was buried in his own section with the exception of the fourth man, and it was six weeks before his body was recovered, and it was found at a place called Kilbride in South Uist, about a mile from Polacharra, the place they started. And his remains were taken to Barra and also buried in the same cemetery where his forefathers were. In the history of Barra, now, I am safe to say that the like of this drowning never before was heard of, and never since did happen.

Tràigh iais

Once upon a time there was found on Tràigh Iais the remains of a beautiful lady. A long period passed and after numerous enquiries it was discovered that it was the body of the daughter of the King of Norway, who last saw her alive whilst bathing on a beautiful beach in that country.

One story is that she is buried beside the schoolhouse at Eoligarry, and that her father sent out a party to erect a memorial and you could see the memorial there to this day. This is, of course, a traditional story which nobody could say whether it is right or wrong.

★   ★   ★

About the Tràigh Iais I am proud to say that it is the finest beach for professional bathers, or any other bathers, in the whole universe.

The softness of the sand, where you can walk with safety – it is as soft as velvet. If you were, on the other hand, a capable surf swimmer I have seen people surf bathing who have been round and round the world, on the various beaches of Africa, Australia and New Zealand, and they came .to a decision that not in any part of the universe did they bathe on a beach equal to Tràigh Iais.

Ecclesiastical traditions

Saint Barr

St Barr was one of the followers of St Columba and he came to Barra and found that we were all heathens, so much so that a missionary came before him and the Barramen got hold of him and they kindled a fire and roasted him and ate him. But St Barr did not meet this fate. He began to go round the island trying to select a spot where to build a church and, going north, it was said that he found a place on the shore of the Tràigh Mhór not far from the stance where the Eoligarry school is today. Then he went still farther north and after going round Kilbar he decided on building the church on Cnoc Chillebarra, where the remains are still to be seen.

He got on very well among his people and they truly loved him and obeyed him, and were very faithful in every way, and particularly as far as the Christian faith was concerned. For many years it was a monastery there, more or less, and priests and bishops were ordained in the church at Kilbar.

When his toil in Barra was over and the day came that he had to leave, he, along with the whole of the islanders went on their knees and he prayed that God would always protect and always support his flock. He gave his blessing to every individual that was standing on the soil of the island, and then to every beast that was standing on the island, and he prayed hard that the sources of the sea would be plentiful, and he blessed the very rocks. And he finished by saying that the last wish that he was praying to God was that the light which he had kindled on Barra would never go out. (And from another story I have told it appears he got a hearing – the day that MacNeil thought the whole island would walk out after him.)

After he left, there was a statue of the saint put up not far from the church, and going in you had to pass round it *deiseil,* sunwise. One night the statue was mysteriously removed and never seen again.

Barramen of that day used to celebrate a day in honour of St Barr and it was mostly spent in shinty-matching, horse-racing, jumping and so on. And none turned a sod of Barra soil the day they were holding a feast in honour of St Barr. Up to maybe a hundred years ago this custom was carried on. Even today people who have friends buried at Kilbar keep up that part of the custom that they do not do any tilling of the ground on that day.

[*See Introduction for an account of St Barr or Findbarr of Cork. His day is on 25th September. His image, which is said to have been covered with a linen shirt on that day, had been removed from the church before the time of the Old Statistical Account (1794). Martin Martin, who did not see it himself, says that the statue stood on the altar; but statues of saints are not kept on altars in Catholic churches. Fr Cornelius Ward says that the statue was in the church at Kilbar when he visited Barra in 1625, although the church was then roofless. The roof would have been of thatch. Fr Ward appears to have got the impression that St Barr himself was buried at Kilbar, but I have not heard any tradition to this effect.*

The horse races and other celebrations connected with St Barr's day were still going on as late as the time of the New Statistical Account (1840) but were tending to die out, probably owing to emigration, the growing poverty of the people and the influx of Protestants into Barra which had taken place during the ownership of General MacNeil (between 1821 and 1831 the population of Barra had fallen from 2303 to 2097, while between 1813 and 1840 the number of Protestants, all incomers, had increased from 60 to 380. See Book of Barra, *p. 179). The custom probably died at the time of the evictions, when misery was widespread.*

Like many local saints in Celtic countries, whose reputation for sanctity and strength of character has endured in the memory of the conservative race for many hundreds of years, St Barr, who of course lived long before the process of canonisation was systematised, is not actually upon the Roman Calendar.]

Saint Brendan

St Brendan was the Sailor Saint, and he did not stay so long in the Island of Barra as St Barr did. And though it is difficult today to believe it, he went out west as far as St Kilda in a coracle. He built a little chapel down at the seaside at Borve and it is called St.Brendan's Church to this day, and the Barramen who live on the southern end of the island keep a holiday in honour of St Brendan and they too do not do any tilling of the ground on that day – it is in May, about the 15th. However, one particular man whom I remember myself – he had no belief in sitting at home and doing nothing on the day of the saint and started to plough in the morning. And what happened – not a grain grew, and the ploughing land was there and never yet did any grow, and never again on St Brendan's day did he turn the ground or plough after that. And this man, he was disliked for doing such a thing, and the day is called yet *Latha Murchaidh Bhig* – Little Murdoch's Day, for the reason nothing ever grew on the ploughing he did on St Brendan's Day.

[*St Brendan may have reached Iceland and the Faeroes in his coracle, as well as St Kilda. For a rationalisation of his voyages, see Vilhjalmur Stefansson's* Great Explorations. *For the life of St Brendan, see C. Plummer's* Lives of the Irish Saints.]

Father Dugan

After the Reformation, when priests were not allowed to land on the island at all, Barra suffered to a great extent, so much so that they had to go to hear Mass in very peculiar corners. One corner I can describe to you is on the Rubha Mór not far from the village of Brevig. On the point of the road they used to meet and on Sunday attend the Mass service, and then disperse, each and every one going to his own home. On several other corners of the island the same was the case.

Latterly the priests were persecuted and not allowed to come to the island at all. That continued for many, many years, with the result that in the evenings people would collect together and say the Rosary. And even that itself faded out at last, and there were a lot of people on the island who were never baptized and a lot never married and still the population was increasing! And that was going on until one day it was learned that Father Dugan came over from Ireland in a coracle.

Well, no doubt the Faith faded out, but the feeling was there the whole time, and when it became known that Fr Dugan was a priest, he was surrounded by all the islanders and he went round and said Mass. Shortly afterwards, you could see the Faith spring up like a hill of heather on fire, and from all parts of the island would come people to the meeting houses (*taighean pubaill*). At one time Father Dugan said Mass right up on top of the hill between Castlebay and Borve, and the place where he said the Mass is still called *Bealach Ui Dhúgáin.*

Now he started to baptize a lot of people who were not baptized; he started to marry a lot of people who were not married, and so on. Now this was going on and then he had to go away, and he went to South Uist and from South Uist to Benbecula, and then he had to cross to the mainland, and whether he ever did go back to Ireland I am not in a position to say. But the day he was leaving Barra, when he announced he was departing, they all mustered round him and kissed his hands and his feet and urgently prayed to him to come back. 'Oh,' he says, 'I would never go, I would never leave Barra,' he says, 'if I was allowed to remain.'

Now he was at Benbecula, and when he had been there some time he had to leave, and half the island only was Catholic. And to this day half the island of Benbecula are Catholics and the other half are Protestants.

And after that, another priest, Fr Fanning, came to Barra and he had no difficulty – the times were not so very pressing as they were when Father Dugan was there. Times were improved and the Faith was beginning to flower out again on the Island of Barra. And so the prophecy was fulfilled that St Barr made about Barra when he said that the fire of faith which he kindled in Barra, he was praying to God it would never go out.

* * *

Many years passed and now we go back and see the day when the Faith is very prosperous in the island. Only MacNeil, who was in the Army during the Napoleonic Wars, said, 'I am going to become a Protestant because I will never get on without.' And he became a Protestant. And he came home, and a few Sundays after he came he suggested to himself he would cause a row between himself and the priest. So he says to the priest, 'I am going to church to-morrow and I am going to walk out (it was Sunday), and you will see that everybody in the church will walk out after me.'

The priest replied, 'I am also going to the church, and,' he says, 'I don't know and I don't care whether you will walk out or not, but I am going to the altar to do my duty.'

So in the middle of the service MacNeil boldly walked out. But not a single soul that was on their knees made a move to follow him, and MacNeil went out very much disappointed, and he never came back to the church any more.

[*The mission of Fr Dugan to the Hebrides is described by Louis Abelli in* La Vie de St Vincent de Paul, *Book I, Chapter xlvi, p. 224. Needless to say, Fr Dugan did not come to Barra from Ireland in a coracle. He was sent by St Vincent de Paul to minister to the spiritually starved Catholics of the Western Highlands and Islands in 1651, and travelled in a ship from Holland to Scotland disguised as a merchant. He was an extremely zealous, conscientious and hard-working missionary in the territories of MacNeil of Barra, Clanranald, MacLeod of Dunvegan and MacDonell of Glengarry and, except in Skye, his mission was a permanent success; but it could not have been nearly so easy if the ground had not been prepared by the labours of the Irish Franciscan missionaries, Frs. Ward and Heggarty in the same districts between 1624 and 1640, labour to which Fr Dugan himself gave rather scanty recognition. Fr Dugan died in South Uist on 17th May 1657, on the eve of what seems to have been a projected visit to St Kilda. (Pabbay is the place named but neither Pabbay south of Barra, nor Pabbay in the Sound of Harris fits the description given, whereas the latter was the residence of the Steward of St Kilda, at any rate not long afterwards in Martin Martin's time). Fr Dugan never realised his hope of visiting North Uist, where he might have repeated the success of Fr Ward in 1625. The reason for the mixture of faiths in Benbecula is largely Protestant immigration.*

It cannot have been the MacNeil of Barra who was in the army in the Napoleonic Wars who walked out of the Catholic Church on Barra in the manner described. He was brought up as a Protestant in any case.

I have been told that it was Roderick, the heir to MacNeil of Barra, who was killed at the seige of Quebec, who left the Church. He must have become a Protestant before he could get a commission in the British Army at that time. It is said on Barra that when the news of his death reached the island, the people proposed to his father, the Chief, that a dance arranged for that day should be cancelled. 'Let the dance go on,' said the old Chief, 'he has dishonoured me in life, and I shall not honour him in death.']

Place names

Bogach na faladh

'The Bog of the Blood'
There was a battle between the Norwegians and the MacNeils on Bogach na Faladh. It was a big battle and the Norwegians had to retreat, and never again did they come back to put up a battle on the shore of Barra, although they had several during the period of their invasion, including the last one on Fuday which I have described to you in another story.

Cnoc a' chrochadair

The 'Hangman's Knoll' on the north side of Bruernish, and it is always called, and will ever be called Cnoc a' Chrochadair. When Mac-Neil of Barra had somebody to hang he would notify the *crochadair* to come up, and his salary was, I understand, this: he had a little dominion there at Cnoc a' Chrochadair and it belonged to him. Hanging was his only job, and he got as salary a free croft for doing it.

Creag gòraich

In the evening, it was customary for the women to go out with pails and buckets of wood to milk the cows, and the fairies or *sithichean* were very common. And when this particular woman started to milk her cow, the *creag* beside her opened up, and from that *creag* came the most beautiful music and song she ever heard. So, the unfortunate part of the story, I cannot find any trace of the song. (If it so happens I will discover it, it will be made known to you.) But it was a very common song at one time on Barra, and the lady before finishing milking the cow had learned it from the fairies.

Many of the songs that were lost in the last sixty years were taken from the fairies – especially when the women of the day were milking the cows in the evening. The rocks or the knoll would open and the finest music that was ever known came from the rocks and the knolls. This class of music was far superior to anything else on the island and even today traces of the fairy music remain in the singing of the *Orain Luadhaidh* (Waulking Songs) and you can hear the difference.

Gleann dorcha

'The Dark Glen'

This man I have mentioned already on more than one occasion, Farquhar MacRae, he used to go to the Gleann Dorcha late at night and see if he could manage to have the pleasure – if you would call it a pleasure – to meet any ghosts. Unfortunately for Farquhar, he never met any!

There is a story of a ghost that was seen many years before Farquhar's time. There was a ship wrecked at Cliat and only two sailors came out. They walked together till they reached the market stance by the crossroads and then one man took the road to Northbay and the other went the road to Gleann Dorcha. And they both perished. And the one that perished in the Gleann Dorcha, it is said that his ghost was often seen where his remains were found. And his ghost was also seen many years before it happened.

Port chula dhubhghaill

The only information I can give regarding Port Chula Dhubhghaill is that there was a man there who lived very near the Port. He was a crofter and his name was Dougall. He was the only tenant at the Port there and hence the name (which means 'The Cove of Dougall's boat.')

Tobar nan ceann on fuday

'The Well of the Heads'

The Barramen and the Norwegians had the last battle and they met on the Island of Fuday – that is an island lying in the Sound of Barra between Barra and South Uist. And I will describe the island first as one of the most charming of the Outer Isles from the Butt of Lewis to Barra Head. And they had a big and a rough battle which ended in a loss of life of the three last Norwegians. And after killing them they were cruel enough to take off their heads and throw them into a well which was nearby. I know the well myself and in fact took buckets of water out of it. On the last day I visited the island I discovered that the well had closed and the water was escaping to the sea through another channel and you could not get a drop of water from it that day.

I remember clearly in the dead of summer you could see the water bubbling up there, but the mysterious point, what happened when it changed its course to the shore, I am not in a position to describe.

So the Norwegians' *dùn* was left unoccupied that day and it has been the same ever since – the last three heroes were killed by the Barramen.

There was a hurricane one year which blew a lot of sand away on the west and north-west of Fuday, and during the storm there were exposed a whole lot of graves, stone-lined, and the skeletons could be seen. And in the morning, when the shepherd saw that, he hurriedly went home and took two of his daughters to help him and his wife to fill up with sand and close the graves, and put them back as near the original positions as possible. And during his long life on Fuday, the shepherd never saw anything like that storm.

Arms on fuday

Now as regards arms which it was rumoured that they were hidden on or after the '45, the shepherd, who knew the whole history of the island, never mentioned to me any such matter.

Tales of treasure

The Dutch ship wrecked off Ard Greian

Once upon a time a ship was wrecked north of Ard Greian, the most westerly point of Barra if not of the British Isles. The ship was coming back from abroad with a cargo of bars of gold, a Dutch ship she was, and there was a tremendous hurricane in the Atlantic and it was do or die for her and she made for the shore of Barra, and she could not make for a worse shore with the wind that was blowing and the sea that was running. And the ship became a total wreck on the reef of Mollachdag – that means in Gaelic 'the cursed rock' – unfortunately with the loss of all hands.

Now a number of years passed and very little of the ship was saved and that same winter put her absolutely to pieces. And very little of the cargo or anything at all was saved – it being bars of gold, you understand, there was no salvation for it but to go to the bottom!

Many years after, there were three women gathering whelks at a very low spring tide and they found bars of what they said to themselves was copper. And they took the copper bars home with them and they decided to show them to an Irishman who then was the only merchant on the island.

The Irishman said to himself, 'Well, I don't believe that this is copper – this may be gold, and so I'd better go to Glasgow myself in the smack that is going (there were no steamers in those days), and take the bars with me and have them examined in my own presence, and see myself what is happening.'

Now he went to Glasgow and consulted the man who was going to identify what was in his possession. And the man started to identify and examine, and, 'What do you say,' the Irishman says, 'the bars are? I took the bars out here to you for you to identify them.'

'Well,' he says, 'the bars are gold and you found them on the

86

bottom of the sea – the gold shows it has been lying in the sea for a long time.'

'I did,' said the Irishman, 'I found them at the bottom of the sea.'

Then he weighed the lot and he told him what was the value and the Irishman had to take him at his word whether he was getting an honest or a dishonest deal. And I am sorry to tell you that I have not the foggiest idea how much he got for the gold, but he bought shawls, blouses and hoods for the ladies, three of each, bought them a box of presents – and I am sure he could well afford to.

On arriving in Barra, he called the ladies who got the bars of gold. 'I am very pleased,' he says, 'that I was very lucky with your copper bars, which I sold, and here is a present for you, Mary – that's a shawl, *a currachd*, that's a blouse and skirt. In case,' he says, 'you would be jealous of one another, I bought the same for the three of you.'

And so the ladies were very pleased at his coming home with such beautiful presents. The dress of the day was two shawls – a small one on the head and a big one round the shoulders. And they all thanked him very sincerely for his very kind gifts, and he told them always to be watching the tide and try to find some more. And they made several attempts to find more of the 'copper' bars but they met no luck, although they got plenty whelks.

Now, after that, the Irishman thrived well, and he put up a big house and he was dealing a lot with cattle. Well, one time he bought two cargoes of cattle and shipped them to the mainland; and he had a son called John and he sent John with them and John sold the lot and over and above squandered the whole issue and asked his father to send him more. And the poor man was sending more and trying to get him back home, until the fortune was exhausted and when the fortune was gone – what happened, John went to sea and became a sailor.

Now he had been away from Barra many, many years and the Irishman died and to my knowledge he had nothing at the latter end.

★   ★   ★

Now I was on drill in Greenock in 1905 and I came across an old sailor and he asked me what part of the world I came from and I told him I came from the Island of Barra. And he says to me, 'Did you ever hear,' he says, 'of a man named John M'G — in Barra?'

'Well, no,' says I, 'I don't know and I never heard of him but I heard of a man M'G — who was a merchant.'

'Well, that's his father,' he says, 'because he told me he was a merchant. And I was shipmates with John – one of the best sailors ever I came across, although he was very foolish for himself.'

In the evening of his days John came home to Barra, as he could sail no longer owing to old age and infirmity. He started to buy sheepskins and rabbit skins and we had a pile just in front of the shop, just ready to make a deal. Well, at that moment I saw an Army officer landing down below the house and I could not identify who he was until he came very close. And who was it but the late Reverend [Monsignor] Canon John MacNeill, formerly in Morar, who was coming home on leave. Well, I was so pleased to see the Canon that I cancelled the bargain I was making with John and told him, 'Take them away, John, I will see you again about the prices.' So I took in Canon MacNeill. He was a co-age of myself and I entertained him as well as the times of the day could afford. And when he was after telling me many stories of the 1914 war he told me a story about Donald Johnston. He was killed and he gave him the last rites, and he had every praise of Johnston for being a brave man and a strong man, and a hero, and he was proud to be a Barraman. He returned and went to Castlebay, the Canon, to see more of his friends, and he called on me before he returned to Eriskay to go back to France again.

One day I was in Castlebay and it happened to be a very hot day, very sunny. Well, I met John on the street in Castlebay and I said to him, 'Mr M'G – ,' says I, 'I have not seen you for a considerable time – what about the price of the skins – the sheepskins and the rabbit skins?'

And he put his hand up above his brows as a shade between him and the sun, and he said, 'I don't know you.'

And that is the way Mr M'G — paid the Coddy for the skins – he never did pay!

The tacksman of Sandray and the crock of gold

Once upon a time there lived a MacNeil on the Island of Sandray and he owned the island. He was also a fisherman and an all-round hard-working man, between fishing and working the Island of

Sandray, and he didn't want from any angle of the compass. During his stay on Sandray a ship was wrecked with a cargo of indigo not far from the Currachan.[1] MacNeil discovered that the cargo was a valuable one, so he started to carry cargoes of it to Greenock, and he did this for a long time without anybody knowing that he was doing anything at all. For each cargo he took out he was paid in gold. Now he was carrying the gold with him home and put it in a peck measure. And he continued doing this for a considerable period, and the peck was getting well-nigh filled. Then it was discovered by the Customs that there was such a thing on the Currachan as this cargo and they came to rescue it, and so MacNeil was knocked out of it. At the same time, MacNeil had kept his hold until his keg of gold was full to overflowing.

So hard-working he was that he wore a shirt going to the fishing that was knotted up nine knots in it, and he had a sheepskin instead of an oilskin whilst hauling the lines.

He then made up his mind to retire and come ashore to Barra. And his choice was Beinn Sgurrival, being the most fertile bit of ground in Eoligarry.

Now, he is there looking well after himself and at the same time getting old. In the evening of his days he made up his mind to marry, and he married a young lassie. He was always a very economical person and he would not go near the keg of gold, which was overflowing, except under very needful circumstances. He had the key of the girnal where the crock of gold was locked safely. It was the custom that a girnal was in every house at the time, one of those big chests about half the size of the room here. He was always manufacturing something, and this particular time he made for himself a pair of shoes of bent grass, which he was using to go out with when the days began to get fine in the spring-time.

Times were going on and the wife was managing more or less about the farm. 'Fear Sgurrival' went out one fine day in April and it was the first day to use the slippers which he had made in the winter-time. Unfortunately, he forgot when he was getting up in the morning to put the key round his neck as usual. And the first thing he did when he got up was to put on the slippers. And when he got them on he walked at his leisure, with his stick in his hand, to the end of the house, where he would be exposed to the sun.

[1] A rock off Bruernish.

Well, he was there now, feeling very comfortable, and suddenly it crossed his mind whether he had put the key round his neck as usual, and he found that he had not put it there at all. And, so as to make sure, he got on his feet immediately and crept silently with the slippers into the house and down to the room. And to his great astonishment he found his young wife up to her eyes in the girnal and in the crock of gold. And he crept very quietly up behind her and took hold of her by the wrist.

'Eh, my dear lassie,' he says, 'you have got plenty.' So she withdrew from the girnal and I don't know how much she had.

But he locked the girnal and sent for the minister, the priest and two elders and he made his will. And the following is the will he made: 'As long as a black cow gives milk and the breakers of the Atlantic ocean break on the shores of Barra, I am giving you this money to see that in my absence, when I am dead, the interest on it will be equally divided among the poor of Barra.'

And that was done. It is a good habit that died out. Information about the present circumstances I am not able to tell. After MacNeil of Sandray's death, his widow lived comfortably at Sgurrival, though she never married again.

[In the Old Statistical Account, 1794, it is stated that: 'the number of poor is generally from 40 to 50; there are £400 sterling of a fund for them, £200 of which is a mortification by Archibald Macneil, late tacksman of Sanndray, and £100 by Roderick Shaw, tacksman of Alasdale, now living; they never go anywhere else to collect their subsistence.'

So presumably MacNeil of Sandray lived about the middle of the eighteenth century. Tacksmen like himself were usually younger sons of the proprietor, who were provided for by being given a 'tack' of part of the estate, sometimes for life, sometimes for the lifetime of the succeeding proprietor, sometimes for a definite period of years. The tendency was, as time proceeded, for old tacks to be superseded by new ones in favour of younger sons of the proprietor of the next generation, so that eventually the descendants of tacksmen were likely to merge with the people in the same way that descendants of the younger sons of peers revert to commoners. but there were exceptions to this, especially in the case of the MacNeils of Vatersay, a family which certainly maintained itself all through the eighteenth century. This family may well have had a claim to the estate, as descended from the dispossessed legal heir Niall Uibhisteach.

The Deed of Entail executed by Colonel Roderick MacNeil of Barra in 1806 sought to limit such tacks to nineteen years or the life of the tacksman, rent to be fixed by auction, but without any grassum.

All the Barra tacksmen emigrated to North America between 1780 and 1830, except the MacNeils of Vatersay, whose last representatives lived in Oban during the latter part of the nineteenth century. At various times there were tacksmen in Brevig, Earsary, Vaslan, Allasdale, Greian, Sandray and Fuday, and perhaps in other places. The list of Hebridean Catholics drawn up in 1703 shows tacksmen, MacNeils in Vatersay, Tangusdale, Greian and Vaslan, and Shaws at Craigstone.]

Tales of local characters

Alexander Ferguson the drover

Back in 1905 I came across a man by the name of Ferguson in
Greenock. His daughter was married to a publican, and I was in
the public house with a friend of mine. And he would not allow me
to stay anywhere else while I was on drill in the naval barracks at
Greenock but with himself. In the said house I met the famous man
Alexander Ferguson, a big man who stood over six feet, and in his
younger days was a cattle dealer. Those were the days when all the
cattle were driven to Falkirk and to all the markets by foot. On hearing I was a Barraman, Alastair got into the news with me very much.

He said he bought cattle at Loch an Dùin and he ferried them at
Northbay and took them over to Uist. And he didn't fail to tell me that
he called in the North Bay Inn, a little inn on the road from the Loch an
Dùin to North Bay. 'At the time,' he says, 'it was owned by a man of the
name of William Sinclair.' I heard a lot about William from several people who lived before me, and I listened to Alexander's story very well.

Now, putting his cattle aboard at Northbay, he ferried them across
the Barra Sound – not a comfortable sea journey in bad weather – and
he landed them on Polacharra, South Uist. After buying some more
at South Uist he carried on north to Benbecula, and after buying
some at Benbecula he went across the North Ford and bought some
at the market stances in North Uist. As it so happens there were two
market stances in North Uist and Alastair bought at both the markets, with the result that he had a 'nice puckle,' as he would say, of
good Highlanders landing on the mainland.

Well, every night, when the toil of the day was over, Alastair
would stay in an old inn and he would leave a little boy to whom he
gave a shilling to look after the cattle at night, and they would graze
at the roadside. And at that inn he bought a bottle of whisky – in

those days it cost about two shillings – and half a stone of oatmeal, which cost about a shilling, and that was his food the whole day until he would reach the next stage house, which was Glenfinnan.

And Alexander followed this routine until one day he found himself in the famous Falkirk – plenty Gaelic there in those days, plenty dealers from the north and plenty beasts in transaction, some selling and some not selling. Well, Alastair said to himself that he was not pleased with the prices and so he decided not to sell and he kept them on hand, along with many others, and they walked to Dumfries. Well, Alastair was not pleased with the prices at Dumfries and so he decided to carry on to Carlisle. All the way from Loch an Dùin, Barra, to Carlisle he went, and Alastair sold there. When he had sold, he got the money in pounds and there was a long way to travel from Carlisle to Muirlagan, Lochaber.

'There were a lot of robbers in them days, and,' Alastair said, 'I had to line all the money and sew it in my jacket, except what I would use on the road. Well, I set off for Lochaber with a good stick and with a good bundle of pounds, from being very successful in Carlisle. And I was taking care of myself and being very careful that no one would take me out of the way, or that I would be drinking with anyone, and I avoided all bad company on the road home in case anything would happen to myself or the money.' So Alastair arrived at Lochaber and no one challenged him.

*   *   *

Alastair was there and his wife, staying with the daughter, and it was most interesting to hear them talking over their younger days. And at this stage the wife said, 'You took care of the money that day, Alastair, but there were many days you didn't do it.' And Alastair replied, 'Whatever I did,' he said, 'you were never hungry since I married you!'

I was very pleased indeed to meet Alastair and Alastair was very pleased to meet myself, and next year when I went to drill in the same place and went to the same house Alastair was very much alive but unfortunately his wife was dead. Before embarking in the steamer in Castlebay I bought a ling from a fisherman I knew, and the ling was six foot four inches. So that I would have it packed properly one of the boys opened it, took the bone out, took the head off and left me just with the fish, and when I arrived in Greenock I stretched the fish out on the table – and the table just could do it and a little more – and Alastair, who was blind, had to feel it with his hands. And he passed a remark that the fish was the biggest ever he had seen – and he was not seeing it at all!

Well, very much enjoying Alastair's company every time I had an opportunity to hear his old news and grand stories, to me very interesting, I was exceedingly sorry the day I parted with Alastair, and not to see him any more.

[*Many stories used to be told in the Highlands and Islands of adventures in the old droving days when cattle were taken south to the big markets at Falkirk.*]

John's sail to Mingulay

John was a very notable person in the Island of Barra and we were both in school together. Then the day came when we parted, and John was to learn to be what was called in those days a pupil teacher and I went to sea. John followed this education and his brains were the best I knew, of superior quality, and he got an opportunity to develop them and he continued being a pupil teacher until he went to college. And then when he got to college he had a lot of interest in navigation; when he was in school with myself he had a lot of interest in it.

One fine day John passed out his exams and he went to teach down Newcastle way, at Gateshead, and he spent a lot of time there as a teacher and then he joined night classes, taking lessons in navigation, still in the hope that he would get to sea some day. Now the time passed and one day he took an exam and got an Extra Master's, as far as navigation was concerned. And now he had to go to sea and learn the seamanship part of the job.

Now what happened but he got schoolmaster, first in the Island of Eigg and then Rum and then Canna, and after that got an appointment and came home to Mingulay. Mingulay is the second last island to Barra Head, where the lighthouse stands, and there are some tremendous cliffs there. And it was John's habit to be climbing the cliffs and getting into marvellous places – one would close his eyes to see the places where John went to the cliffs and stood out on top of them!

Now he had a small boat of his own and he used to go to Castlebay with it all the way from Mingulay, and if he was not going to be back on Mingulay for a day or so, he used to put up a notice at school saying, 'This school will be closed this week-end till Monday' – giving himself plenty of time.

One day he was coming across from Castlebay on a Saturday. There was not a breath of wind in the whole horizon – in the area in which he was, anyway – and he said to himself, 'There is nothing for it, boy. Just take off your jacket and row to Mingulay.' He put off his boots and made himself as light as possible, and then something happened in the interval – and unfortunately what happened was, he fell over the side. Fine and sunny it was and John caught hold of the gunwale but, although he did that, he could not take himself out of the water and into the boat.

'O, vó vó vó – how am I going to get out of this difficulty?' he says. And he was looking around him and making another attempt and still failed. Now the helm and the rudder were out and so he got within reach of the helm, took off the helm, took off the rudder and put them both inside the boat. Now he caught hold by the stern and he put his big toe on the lower iron for holding the rudder for a support, and hove himself into the boat. Well, he was successful and, with the assistance of the iron that was to hold on to the rudder, he gave himself a spring and that brought him into the boat.

Now John was standing aboard and, 'Nothing else for it, boy,' he says, 'plenty sunshine, strip off, be the wild man in the Philippine Islands and row bare in the sun to Mingulay.' And he spread his clothes all over the little boat in the thwarts, and there they were.

That was the position, now, until he came in sight of the beach at Mingulay, and he knew that a lot of people would be waiting on the beach to give him a lift to draw the boat high and dry, and so he got himself respectably clad before meeting the crowd on the beach.

'Well,' he says, 'well, my dear men, I am sorry to tell you, you were very nearly minus a schoolmaster today. When I was off the Island of Sandray,' John said, 'I fell over the side and by good luck, and the hand of Providence being over me, I took off the helm, took off the rudder, and with the assistance of the lower iron that was holding the rudder on, I managed to get aboard.'

And they all wept, you know, and said, 'Oh, John, we are very pleased to see you.' And then he said, 'Get hold of the painter and we will draw the boat ashore.' So when they got her beached, after throwing out the ballast, John jumped out and he was quite dry. Now they were merry, and someone went up and kindled a fire at the schoolhouse there.

[*Mingulay, where John was teaching, is about fifteen miles south of Castlebay, and has no natural anchorage. There used to be eighteen families living there, but they left in 1908 and settled on Vatersay and near Castlebay.*]

More about John

This particular day we were out fishing and it was an exceptionally beautiful day – it was the beauty of the day put me in mind of the story. 'Well, John,' says I, 'we are having a good time.'

'Oh yes, Coddy, we are having a good time. Isn't this now far better than standing teaching a lot of poor little children that the Almighty didn't put any sense in their heads – and here's poor John standing trying to knock sense in, that the Creator failed to do.'

That is the picture he made to me of his job, standing in the boat with the Coddy!

Another day we were going to Eriskay, John and myself. There was an Eriskay man after shooting his lines and he was sitting comfortably in his boat having a little doss to himself.

'By the by, Coddy,' says John, 'who is that man in the boat?'

'That's Angus MacKinnon,' says I.

'Well,' John says, 'he is enjoying the day very much and he is lying down there – I don't believe he could be more happy in Heaven than waiting till he starts to haul his lines.'

So we passed to Eriskay and did not rouse him at all; it would be a crime to destroy his slumber.

Alastair and the tenth of May

One year in Castlebay there was a big dispute between the fishermen of Barra and the curers, and the fishermen were not keen to allow herring to be landed before the 10th May. Now there was a big fleet in Castlebay and they challenged the Barramen that they would go to sea and the curers would buy, so the Barramen said, 'You won't, because the herring is not mature and the time is not ripe for curing and the reputation of the Barra herring on the continental markets we have had from time immemorial, you would be letting that down.' However, the day came on which the decision was given. All the Barramen went up to Castlebay to see that they would not be allowed to land any herring until this day. Now they told the fishermen and the curers they had no ruling from the Fishery Board over them, the natives of the island, and it was the decision of the Barramen that had to be taken. So Alastair was the head of this and they would not allow any herring to be brought in.

And who should be very strong against the Barramen but a black-bearded policeman who was on the island, a Mr Tait, a much disliked human being. And my friend Alastair had no love for him either.

Now Alastair held a conference and they formed a body on the beach where the samples were to be landed– 'And I will preach against you to let the herring go on.'

The first man that came with the sample was a man from the port of Hopeman. And the policeman was there to catch a sample and Alastair formed a ring of Barramen round himself, and he turned, and they began to push Alastair into the sea. Alastair stopped them and said, 'Are you going to put me into the sea?' Alastair was near the policeman by this time and he turned round to the mob and told them off and swore at them: 'Ye children of the Devil, are you going to put me into the sea?' – and the more Alastair cried the more they pressed on him, until he was very near the policeman, and he gave him one kick and threw him into the water.

Immediately the policeman was into the water Alastair said, 'Come on, get on with the job – here's the policeman in the water – get that scummer.' There was a scummer lying in the visibility. Someone handed Alastair the scummer, he put it round the policeman and dragged him quietly ashore and as Alastair said, 'I would give him a kick but it was against the law of the country.' And the policeman walked quietly away and the Barramen got their way of it and Alastair was very proud that his scheme worked out well and that he was the conqueror. And there was no herring caught until the time approved by the Board, and given by the Barramen, and the Barra fishermen won the day.

[Alasdair Aonghais Mhóir was a well-known character on Barra. He was a Gaelic story-teller too.]

Alastair and the pigs

Alastair was one night in the mill, and the miller kept him behind all the others because he lived nearer the mill. And also the mill was not working too well owing to the shortage of water, and that also kept Alastair behind. However, he got his meal into his bags ready to go and the miller gave him a hand putting the sack on the pony. By this time it was getting near the early hours of the morning, and when Alastair was about two hundred yards from the mill, there is a stream running on the north side of the road called Abhainn Gluig. On drawing nearer to it Alastair saw six pigs, and when Alastair saw the six pigs his hair stood on end with fear. Immediately he made a sign of the cross with the water of the stream and then he turned round and threw a good sprinkling on the pigs. Immediately Alastair had thrown the water which he blessed himself on the pigs, 'they scattered,' says Alastair, 'on fire, and went right up to the skies and into the kingdom of Satan – where they came from!' So nobody saw the pigs after Alastair gave them that shower-bath from Abhainn Gluig.

[There were no pigs kept on Barra, then or now.]

The story of the thrush

One particular Saturday Iain was in Castlebay and he fell into a tremendous company which lured him away. And they were turning them out at ten, and he was making a very poor passage, Iain was. He had with him a pint of whisky. Well, coming down about Tangusdale the road was getting the better of Iain, and he was not making what you would call a passage of it at all, and so he sat down. And with the peace and quietness and the beauty of the night, he went to sleep and that carried him to the early hours of the morning, and when he woke he was in such a terrible condition that he vowed by all that was holy that he would never take another dram. Before he was properly awake he thought he was hearing music. Lo and behold, what music was that but a thrush crooning to herself the most beautiful of music, and saying, 'Poor Iain, you-are-very-dry, you-are-very-dry, you-are-very-dry' – 'Iain mac Iain, 's-tu-tha-tioram, 's-tu-tha-tioram, 's-tu-tha-tioram' – several times in succession.

After hearing the thrush continually coming out with the same tune he then gave a great sigh and he said to himself, 'Well, the poor thrush is telling the truth, and even though I promised myself I would not take any more drams, I had better take a drop.'

Then he was beginning to talk to himself, saying this and the next thing, and after he got that life-saver he began to feel a little more comfortable. And all of a sudden he thought that the thrush changed the tune, and he said that the thrush's tune this time was, 'Iain, Iain, take-another-mouthful, take-another-mouthful, take-another-mouthful' – 'Iain mac Iain, gabh-balgam-eile, gabh-balgam-eile, gabh-balgam-eile.'

Which he did.

By then there was not much left in the bottle. As he was looking at the bottle again, Iain heard the thrush begin to sing another tune, 'Iain-finish-it, Iain-finish-it, Iain-finish-it' – 'Iain mac Iain, cuir-crìoch-air, cuir-crìoch-air, cuir-crìoch-air.' And the mouthfuls continued until latterly there was not any more to take – the bottle was dry.

Well, the next stage, he decided that he had better go back to Castlebay anyway, and he said it was when he was already half-way up that he thought he had better turn. And so Iain arrived at the pub one of the earliest customers that day.

[*This story occurs elsewhere. See Holmer,* The Gaelic of Arran, *p. 175.*]

The betrothal and wedding of John the fisherman

John was at first a fisherman and then a sailor, and without a doubt one of the best, if not the best, the island ever reared. John was a widow's son, and it is said that his father got gold on the shore from a coffin which contained a well-to-do lady who was buried at sea, and a crock containing a quantity of gold was put in the coffin along with the lady to bury her. The lady was buried quietly, minus the gold. John, when he grew up, followed the fishing both on the west and east coasts of Scotland.

Once on the east, fishing out of Fraserburgh, he had done a

bumper of a fishing, so when he came home he went without much delay to see his favourite girl. After several nights of this routine John decided to pop the famous and important question, 'Will you marry me, Catriona? – and if you agree we shall do it without any delay.' It took Catriona some time before she replied, 'yes' in a low and humble tone, but if she knew what was lying ahead of her – and very near – her reply would be very doubtful!

However, it was decided that the betrothal would be the very first Friday night. John got ready to go to the nearest inn, which was Pollachar, and ordered seven gallons of whisky at fifteen shillings per gallon, five pounds five shillings – cheap compared with today.

John and Catriona were great favourites on the island, which meant that nearly the whole island had to be invited. This was done. It was a tremendous feast; some hundred hens were killed and four or five sheep. Now the party assembled. The cooks and stewards were cooking outside and the usual ceremony gone through, and when this came to an end, Catriona got her parents' consent.

The eating and drinking and dancing went on at a furious rate – so much so that the fire on the middle of the floor was danced out of existence, and all the ashes went out through the hole in the roof, that was meant for this kind of escape. This went on for a considerable time but unfortunately whisky began, and too much of it began, to speak. Some of the men began to fight and at the latter end they were all fighting, and now there was still showing no signs of peace. There was an outbreak among the cooks and stewards, and at last John's betrothal developed into a hurricane of eating and drinking, dancing and fighting.

This carried on until the cock crew – as soon as that happened peace was declared. John himself, who told me the story, said the Cock frightened the Devil away.

The party separated on Saturday morning. All day Saturday the news was going round about the night before and John was not too happy about it, getting all the blame for being so generous about the whisky. On Sunday morning John had to go to Daliburgh to put in the proclamations. When he came in sight of Lochboisdale he saw a steamer lying at the pier. Instead of going to the church he made a bee-line for Lochboisdale and was twenty-one years before he returned!

★ ★ ★

Now his adventures through all those years were very numerous – in fact, too numerous to mention. John joined the steamer at Lochboisdale, calling in various other ports on the way to Glasgow. The voyage took ten days. The first move John made was to go down to the docks and admire the sailing ships. He very much admired one in particular, catching the eye of the mate. He asked John did he want a ship?

John replied, 'Yes, I do.'

'Where are your discharges?'

John replied, 'I have not got any.'

'And in what capacity were you going to sea?'

John replied, 'I was a fisherman.'

'Very good, my young man. My father was a fisherman, and also a good seaman.'

John joined the ship without any delay. Never having seen a sailing ship before, he began to acquaint himself with sails and all other jobs on board the ship. One day they were ordered aloft – John was rather stiff the first day! When they got to sea it was all plain sailing to him.

The first voyage was to New York. If he was going to New York today he would find a mighty difference. At the end of every voyage John would have a few days' celebrations, promising himself, 'I am going back to Eriskay and marry Catriona.' This routine continued for many a long year.

John was promoted to the capacity of a boatswain. One day they were lying at anchor in the River Amazon. Through the captain's fault John and the second mate fell out and the fight started and John in one of the rounds knocked out for ever the mate. Now the captain was called. The mate was declared dead and the verdict against John was that he was to be hanged to the yardarm tomorrow. On the top of this he had to keep the anchor watch to night. This was to be his last night on the face of this earth.

John went on duty. First he began to leave it to luck that he might be forgiven. Latterly he decided on swimming ashore and trusting to providence. At the dead of night, when all was still and silent, John made up his haversack and tied it well on his back. Before taking the plunge he went on his knees and prayed and promised if God would spare him to get ashore this time not another voyage would he ever make but one for Scotland, and see Catriona and marry her.

Now he takes the plunge, and to swim maybe the last mile in his life. He made a beautiful picture of hope that he would see Catriona. When he was coming to the end of the mile he was feeling tired but still he had sufficient energy to walk up safely on the rocks and thanked God for his safety. Catriona again came into the picture and a second vow of seeing her sure this time passed through his heart most affectionately. All of a sudden a tremendous shark jumped out of the water, so near to the shore that some of the spray hit John on the head while he was yet dressing, and a third vow he made to see Catriona and to marry her.

The road was just near. All was still and silent. After going up the road he heard a furious noise in one of the buildings, so furious that he thought it was an asylum! On making a latter survey he found it out to be a lot of people rolling on iron rollers – that is exactly the term he used. He continued to walk until he was a considerable distance from the port. Feeling hungry, he went to a farmer's house and asked for some food. This he did get. He then asked the farmer what hopes there were of a job in this locality. The farmer said there was a coal-mine – people were always coming and going from there. John went over to the coal-mine and got a job and stayed for over two years. Then he went to look for a ship and with his earnings in the coal-mine and his homeward voyage, he counted on going home to Eriskay with a handsome sum to marry Catriona, which had to come before any other enjoyment.

John arrived, and got to Eriskay and made a bee-line to see Catriona, who did not know he was coming home. Since John left, Father Allan McDonald was the priest in charge of Eriskay.[1] John went to see him and told him his story, yes, his wonderful story.

Next Sunday John was to be proclaimed after twenty-one years' absence. Immediately after it became known that John was getting married, the boys started to build mounds of old boots, tin cans and everything that could be thrown except stones. John went to see Father Allan and told him what was happening, so they decided to come to the church in a boat, Catriona, himself, the best man and the bridesmaid, and so that evening saw John and Catriona married.

★　★　★

John is now married and settled down. One day he said to

---

[1] Fr Allan McDonald was priest on Eriskay from 1893 to 1905.

Catriona, 'What about going on a little honeymoon?'

'And what is that, John dear?'

'It is a usual trip taken by people who marry,' said John.

They invited a few friends with them and then they arranged to go and take John's own boat. Little did poor Catriona know that John was planning a drinking expedition. However, they sailed from Eriskay to Lochboisdale. When they left, the weather was fine, but towards the evening the weather got stormy, in more ways than one. John came down from the hotel very much under the weather, so much so that Catriona refused to go into the boat with him, but she went into the small boat. The wind was blowing a Sou'-sou'-east dead ahead of Eriskay. When they went out a considerable distance in the Minch they put about and Catriona was speaking to them.

Now they were on the homeward tack. They had another dive at the bottles and when they got on deck looking out for the Haun at Eriskay they discovered that the small boat was adrift. Then there was a terrible panic on board, tacking and turning all the night. They went to the forecastle and said their prayers – John was in a terrible stupor, and he said that Catriona was dead, and without a doubt she was in her mother's lap in Heaven.

Now it was getting daylight. One of the boys stood forward and saw a boat on the top of a rock called Sgeir an Fhéidh (the Deer's Skerry). Coming closer, they saw Catriona, and with very good skill they got the boat alongside the rock and took Catriona and the boat out of the difficulty. John took her in his arms and carried her in the forecastle, put on her some of his own dry clothes, and gave her tea and some whisky – which she did not want to take. In a short time she was all right and none the worse for her adventure.

When they were passing Lochboisdale John wanted to go ashore to have another burst to celebrate that they had got Catriona back again.

'Some other day, John,' said Catriona.

John the fisherman's Christmas homecoming in a blizzard

At the beginning of November it was customary to go to the lochs, Loch Hourn and Loch Nevis, to fish herrings. Now there were no clocks or watches in those days. John had to take with him a cock so that he would be crowing in the morning. The cock was very useful every day except Sunday, and John did not want to hear him on Sunday as, being awake the whole week, he was keen to have a long sleep. And so annoying was the cock to John on Sunday morning that he was feeling inclined on many occasions to get up and twist his neck. However, the time went on and the poor cock escaped the penalty of death. Now the season was over and it was getting near Christmas time and there were three of the boat's crew married men and they were all keen to get home for Christmas and John was the only single man on board.

At cockcrow, when they decided to leave for Eriskay, they got everything in readiness for the long sail before them. Now the wind was a nor'easter and a very fair wind and promising snow, and when they came abreast of the Island of Canna there was a great conference to decide finally whether to let her cross to Eriskay or not. Well, John was independent whether he would go or not and he was leaving it for the older men to decide, and then the three others were keen to go. And the wind was getting up and the night was coming on, and no compass, don't forget and so they let her go.

And when they were about an hour or so after passing Canna out of sight in the Blizzard, they began to tremble. The night was passing, no sign of land, and they just had to steer – they did not know where. Well, one began to cry and to weep and wail and latterly they all began to cry, for their wives and their children, thinking that their doom was at hand. All John said when he was seeing that they were tired crying, 'Well, boys,' he says, 'we are not going to die yet. I just have a feeling – my upper lip is very itchy, and that never happened to me but a dram was not far away.'

And they turned round and said to him, 'You should not be speaking like that, John, in weather like this. You have no sense.'

'Well, I am telling you, Donald,' he says 'that I never had in all my life the itchiness of the upper lip but there was a dram near it.'

And between the intervals John was cheering them up, keeping well before them that the dram was near.

At one time in the middle of the argument John felt his lip getting itchier and itchier and, 'Oh,' he says, 'the land is getting near.'

Now there was a pause and nobody saying anything, and John had the third turn of the itch, and he jumped up and went to the bow of the boat and without a shadow of a doubt he was quite clear that he was seeing the point of Flodday, an island at the back of

Fuiay there. He said, 'Here is the first buoy – this is Flodday Island.' And he had another itch and he says, 'I told you, Donald, I was never let down' – and the wind backing from the north-east, they made shelter and they followed the. way in right to the top of North Bay here, just below the priest's house. And they moored the boat and landed there and went to the old inn, along with Mr Sinclair. And Sinclair was astonished to see them and didn't know where they came from in weather like this.

It happened that he had in the house an old piper, called Donald. I have seen old Donald myself, and he loved the dram, did Donald. And he never refused a dram when he was at a wedding – when he was full himself he took the dram and blew it down the pipes. John ordered a bottle – and a big one – but out of the hundred pounds they made at the fishing this was little enough. They started the bottle which they punished severely and in very quick time and gave Donald a good share. Then the stories began and another bottle was called. Then John felt inclined to have a dance for himself and told Donald to play the pipes. The four boys stood up and danced a reel just as they were, in their oilskins and seaboots. When they were finished they put off their oilskins, and then John says, 'Ach, you devils,' he says, 'didn't I tell you, Donald, that the itchiness of my upper lip was ever and was always a good forecast that a dram was near at hand?'

So the old woman of the house, Mrs. Sinclair, at that stage gave them something to eat, and they had a good feed and Donald played the pipes and John called another bottle, that is, a third one, and at the third bottle the old woman says, 'Well,' she says, 'that is all you will get.'

'Och,' says John to her, 'my dear lassie, you will have to bring up a gallon jar yet.'

'Oh you will have to do without the jar to-night – this is Saturday night.'

She would not give them any more, but sent them to bed and they were very comfortable. And the next day they saw their wives and children in very good order.

[*When itching was felt it was considered a premonition, e.g.* sgrìob an airgid, *money foretold by itchiness of the palm, and so on.*]

William MacGillivray and the bagpipes found at Culloden

On the field of Culloden Moor one time was found bagpipes, and a bonny set they were. They were taken to Greenock by the man who found them on the field. Now an uncle of the MacGillivrays of Eoligarry was one time out in Greenock. He was a piper and was very interested in the pipes, especially when he heard that they were found on the battlefield of Culloden. And he being a blood relation of the man who they belonged to, MacLennan asked could he have a loan of the pipes – either that or could he buy them?

The man said, 'No, you are a piper and I am not, and however I will have much pleasure in giving you the pipes, as you can play them.'

So he took them with him to Eoligarry and he left them in charge of MacGillivray's two boys who were then learning to be pipers – and I need not confess to you that they were taken care of there – the boys are very fond of playing them and they remained in that house for over a hundred years.

Well, William was the last surviving of the boys and it was always troubling him what would happen to the pipes when he would fade away. And then he decided to give them as a present to the lady who was in charge of the West Highland Museum at Fort William. Her name is Mistress Ryan. I was on one occasion going to Inverness to a County Council meeting and William asked me would I convey the pipes to Lochaber – that he was going to give them to Mistress Ryan of Spean Bridge, who was going to see that they would be well looked after in the West Highland Museum. So a day before I left for Inverness I called and the pipes were made up in a beautiful parcel and sealed more than a few times – I believe it took him a fortnight to do it! – and, 'Here you are,' he said, 'John, here are the pipes and give them to Mistress Ryan, and I sincerely hope that they will be looked after as well as we did for the last hundred years.'

I took my very best care of them. I asked Captain Duncan Robertson if I could put the pipes in his cabin, to make sure nothing would happen to them, and he said, 'Oh yes, certainly, Coddy.' Now I visited him again in his cabin to see if the pipes were in order and then he said, 'What have you got in this wonderful parcel?'

'It is not a bottle anyway, Captain,' I said. And then I told him the story about the pipes and at that he took out a knife to cut the string and open the parcel.

'Oh no. Captain, you are not going to do any such like,' I said. And so I arrived in Lochaber, I had an interview with Mistress Ryan, and gave her the pipes.

★   ★   ★

I went in one time to see what was actually happening to the pipes and I found they were hanging on a wall and looked more or less neglected. And I told the lady, Miss MacGregor, 'Well,' says I, 'I took those pipes from Barra to Lochaber and the man who looked after them for a hundred years would not like to see them there. They are worthy,' says I, 'of being put into a glass case. They were found on the Moor of Culloden.'

I was not very long in calling again and I asked for the pipes, and the man in charge took me to a glass case and here they were lying there, beautiful.

Several years after that the 1938 Exhibition of Glasgow was on, and I met Mistress Ryan and I told her the story of the pipes in full, and I suggested it would be a very good idea to have them played at the Exhibition – a musical instrument which was very important and was very well preserved, and she jumped at the opportunity at once, and asked me who would be the best piper in my estimation to play them, and I said to her I would not leave Lochaber behind me before doing that, and I told her Angus Campbell. Angus was given the pipes so that he would have them in readiness to play when the time came, and he went to the Exhibition with the pipes and it was announced in the B.B.C. *Radio Times* that week that such and such a man, Angus Campbell, was to play the pipes, and the whole history as I have given it to you.

Now MacGillivray of Eoligarry was notified that the pipes would be played by Angus Campbell, and the moment he heard of it he sent a message to the minister to go and collect the Coddy and take him down to Eoligarry, so that he would be with him listening to the pipes once more, and the minister, of course, obeyed the order and he went and called on me and we went down. So the tune that Campbell played was 'MacIntyre's Lament' – a very pathetic pibroch. I was carefully watching the old man, and as Angus Campbell was going on into the heart of the piping I saw poor MacGillivray's eyes getting very moist and I clearly understood what was inside. And when Angus was finished, William was leaning heavily on his stick, his eyes were soaking in tears. He could not for a while give us even a remark. Latterly he says, 'I am very pleased to hear such an excellent player handling them to-night.'

[*The MacGillivrays became tenants of Eoligarry after General Mac-Neil sold the Isle of Barra in 1838.*]

Stories of the *Politician*

Medicine from the 'Polly'

On the island of Eriskay during the *Politician* period, when whisky was running like rivers, one day a crofter had his stirk very, very seriously ill, and in the absence of the vet John suggested to himself that he would tell his neighbour. And when his neighbour arrived, the poor stirk was lying stretched out on the ground, and could not lift his head, or his feet or his tail.

'Well,' his neighbour said, 'if the stirk was mine,' he says, 'I would give him a drop of the *Politician*. At least it would do him no harm, even if it does him not so much good.'

Well, the whisky was plentiful, and they had no distance to look for it, and John got a pint. He took hold of the bottle and opened the stirk's mouth and put his thumb on his tongue so that the whisky would flow down gently. They were anxiously waiting for the results of the medicine, and before long the stirk began to open his eyes and show signs of life, and John told his neighbour, 'Well,' he says, 'I think he is none the worse of it.'

Now after a pause John said it would be a good idea to give him a little drop of hot water and a bit of sugar, and make a toddy for him. So this was arranged – the toddy was made with some water and sugar, and John, who was a capital hand at serving the medicine, gave it to the stirk. Then they let him lie down for another while and shortly after that John reported that his eyes were getting much brighter than they were before. Then the stirk made an attempt to lift his head off the ground. And John remarked this, that the stirk was rapidly improving – and on the strength of his neighbour's suggestion they arranged to give him a third dram.

Well, when he got the third one, he made a wonderful recovery, and he sat up and began to shake himself, and then he bellowed. Then John comes on the scene again, and he says, 'The stirk has made a wonderful recovery and I think if we give him another one it will be a complete cure!'

The same performance was done, that being the fourth dram – and when he got the fourth he really and truly made a pure shake that almost shook his tail off, gave one bellow, and off he went with his tail curled, making for the highest hill in the island. Both men were amazed to see the pace he was going at as he climbed the highest hill, called Beinn Sgrian. And when he got to the very top he was amusing himself there, running and playing about. Then he came down the hill and started to graze, in his usual good health.

So that is one story of a complete cure from the world-famous *Politician*.

[*The* Politician *was a merchant ship carrying a cargo of over thirty of the best brands of whisky to America which ran ashore in 1941 on the small rock called Hartamul near Eriskay and was abandoned. The cargo – or a good part of it – was salvaged by willing rescuers, and did much to enliven the dark days of the war in the Outer Isles. Many songs were made about the incident; one of them is printed in Margaret Fay Shaw's* Folksongs and Folklore of South Uist, *and the event is also the basis of the well-known novel and film,* Whisky Galore, *by Sir Compton Mackenzie.*]

Hiding the 'Polly'

During the run of the *Politician* there lived in the Island of Eriskay a good-hearted Highlander, to name Ronald. He was a fisherman and a very good one. Like many other Eriskay men – and more than Eriskay men – he was a famous hand at the 'Polly.' In fact, he had a tremendous stock in hand when the rumours came that an invasion of Excise officers were on the road to Eriskay. And he said to the wife, 'Well Catriona, we had better get the whisky out of the house,' he says, 'because the Excisemen are coming.'

'Oh, yes,' said Catriona, 'for God's sake take it away and don't let me see a drop of it coming in the house again. I am not getting a wink of sleep since it came into the house whatever, and you should have put it away long ago.'

Now in the twilight this certain night Ronald filled a sugar bag which would contain three or four cases, and he put the whole show on his back, and he made an attempt to move his stock out of the house and hide it in the hill. Ronald was very tired of the bag before he reached his destination, and when he made himself believe that he was in the safest spot in Eriskay for hiding the lot, he let go the bag and emptied it all out. Now next he got hold of a bottle and he tried to hide it into a corner of a rabbit burrow – and lo and behold, what happened, it struck against another one! Well, he tried it not far from the same place, and the same thing happened, and for six occasions in succession Ronald could not get his bottle into a burrow because there was one there before him. So he gave up the ghost and returned to the dump, and he was admiring the bottles there and was going to bid them goodnight for ever, because tomorrow they would belong to someone else.

Then he saw there on the side of the bottle, 'The King's Ransom,' and 'Well I will see no more of those bottles' says Ronald, 'and I think I will take a dram out of the King's Ransom' – so that Ronald did.

The medicine began to feel happy on Ronald, and he decided on taking another one, and at that one he sat down and treated himself to a smoke. And he said, 'Ah, well, I am very disappointed to be going away and leaving all these bottles behind – I think I'll take another dram.' So that was Number Three. And so on, continuing at that rate the bottle was very nearly empty before Ronald said to himself, 'I had better go home,' so he made for the homeward journey. And I am telling you, he was making much better way on the outward journey! And crawling from port to starboard he was in a merry state – or perhaps a poor state.

It was twilight, and the wife could see Ronald coming home and she was wondering what was the matter with him and when he came closer she discovered he was much the worse from the expedition to hide the 'Polly.' And she rolled out in Gaelic, "*Raghnaill, Raghnaill, de tha cearr ort, a m'eudail?*"(Oh, Ronald, Ronald, what is the matter with you, my dear?)

And Ronald said, 'You know, Catriona, I took with me to the hill a big bag and I was certainly dead tired before I got it there. On arriving at the place where I was going to hide the bottles not a damn corner was there but there was a bottle in every burrow before me. And I drank the biggest part of one myself and now, supposing all the Excisemen between this and Hell itself comes, I would not go another inch to Ben Stack' – and Ronald went to bed.

And the next day, when he took a walk to admire the scenery he left yesterday, the coast was swept – there was nothing there.

Transporting the 'Polly'

When the whisky was very plentiful in the Island of Eriskay, some party hired me over with the boat – not to visit the *Polly*, but to visit their friends. It was a very, very hard day for me – so many Eriskay people knew me, and so numerous were the offers to take a dram, that I vowed to myself, 'I am not going to take any.'

Well, then, I was trying to make a deal with a man who had several cases up on a hill. This is how I did the deal. I told him to get four empty bags and put a case in each bag, and it was in the month of May, and the coal and peats were scarce, and I smuggled the peats and the whisky together. I told him to put them into the bags, the whisky first and the peats on top. I told him I would wait until the fall of the night, and when he would see me leaving the Rubha Ban that I would meet him over at Sloc na Creiche and he was going to take the bags down there, and the cases, and he would have the four bags ready for me when I would arrive there.

Now I am aboard. We bid goodbye to my numerous friends who came to see me off and very sorry they were that I would not take some whisky. However, the man with whom I made the deal for the peats was going to await my arrival at Sloc na Creiche, and the place is where the Englishmen killed the Weaver and his three sons.[1] Everything was still and silent and those that had been at the party were well equipped in bottles, and I had none, and when we left we went off our course a considerable distance to get to the Sloc where my client was waiting for me. The peats were there and we hauled aboard the peats, and there was a priest aboard, and nobody was a bit the wiser, but one inquisitive fellow, seeing them, 'Why,' he says, 'are you importing peats to Barra?'

'Oh,' says I, 'the peats are so wet just now and my coal boat has not come, and it is under difficulties, and very severe difficulties, that I had to come today to take four bags of peat to Barra.' I don't know whether he took that with a grain of salt, anyway he made no further comment.

We arrived and everybody went ashore. 'Now,' says I, 'you are all going to leave me and the peats here, and nobody will give me a hand, after me taking you to Eriskay and back!' No sooner was that order piped than the peats were ashore and the order obeyed very quickly.

[1] See page 60.

In a year's time I told the story to the priest, that I had four cases of whisky among the peats which he had seen in the boat the night we came from Eriskay. 'Ah, you rascal, and a great rascal too, I was always wondering why you were going off your course, instead of coming to Barra!' he says. 'You were steering right out into the Minch, to Canna I would say, or Rum.'

Another tale of the Politician

During the raids on the *Politician* one particular peculiar thing happened. One day there were three cases of whisky found in bags not far from where my boat was at anchor. Somebody discovered the bags and came and told me of it, and I said to the fellow, 'Well, my dear boys,' says I, 'I don't know who left them or who didn't leave them there, but I will attend to them meantime, until I find further evidence who did leave them. But thank you for telling me, and don't tell anyone you told me about them.'

About four o'clock in the evening this man came back and asked me what happened to the whisky bags. 'Nothing to my knowledge – I believe they are there yet,' I replied. We looked over and there they were, and when the night closed down the bags were in safe hands!

And this is an example of how the raiding of the *Polly* took place, and those that were not at the *Polly* at all sometimes were better off than those who went through a lot of difficulty acting the raiders. I will tell of an instance of how a man had his share hidden among the peats, and he was quite sure that no one would ever get hold of his lot, but one night he found himself very disappointed. And when the raiding party came and removed all that was there, one of them – a brilliant fellow – wrote a note and put it on one of the peat bags. 'Dear Willy,' it ran, 'I am sorry to tell you that the remains have eloped!' So those that got it in the end were not at the *Polly* at all.

There was a woman down at the North end here and she had some friends at Castlebay. The Castlebay friends wrote her a note asking her to give the two children two bottles of whisky – a bottle each. And she replied to the note, 'My dear Mary,' she said, 'I am sorry I have no whisky today to give you. The last I had, last night I bathed my feet in it, owing to the fact that I was severely bad with the rheumatism. But send down the children to-morrow and if all is well Donald will be at the *Polly* to-night and I am sure to be able to send you some.

Other instances of this description did happen during the reign of the *Polly*.

Stories of sea monsters

The capture of the huge basking shark

In olden times Barra fishermen were famous for fishing the basking sharks and they had a harpoon for the job, and an axe and a huge coil of rope so that, when the crew were set, the harpoon-man was standing in the bow of the boat and he always harpooned the shark head on – he could not go near the shark's tail because if he felt the harpoon was going into him he made a tremendous splash with his tail which would capsize a boat of no small size. And there was a man ready with an axe so that if the rope went foul, or anything went wrong, his job was with one stroke of the axe to cut the rope and the boat was clear. For it happened that some time before the time I am talking about, something went foul and the shark took the boat to the bottom.

One day, off the shore of Mingulay, the men could see a basking shark not far away from the shore. So they mustered the crew and got ready the boat, got hold of the harpooner, and out they went. And with the first blow the man harpooned the shark – and as soon as ever the shark felt that happen, his effort was to go to the bottom at full speed and try to get the harpoon out of him. And he went down and rubbed himself on the bottom, but instead of it coming out, he let his whole weight on to it and it went deeply into him.

Shortly afterwards, the shark came to the surface, and for the rest of the day he kept going round and round and latterly, when the evening was coming on, he made for the Sound of Mingulay, and the people of the island followed round and they saw the shark going far out into the Atlantic.

At sunset they could see the boat and no word of coming back.

Then they wept and returned home thinking that all hope was lost that the boat and the crew would never be seen again. It came late at night and they began to say their prayers. Latterly they had a sleep, and then one man happened to look out on the beach below the village and here was the boat and the shark and the crew! After some time through the night the shark changed its course and came to Mingulay exactly the way it went out, with the result that latterly he got tired and he beached himself on the sands at Mingulay. And the men were delighted to be ashore, and they had no trouble getting their knives ready and operated on the shark and took out of his body several casks of liver, and the catch was very valuable to them.

The sea monster

The story is from my father and from the people who were with him in the boat at the time.

One day my father was on the west side of Barra fishing lobsters and it was in the herring fishing season, and lo and behold, they saw what they thought was a fleet of herring nets that were lost by a boat – it was that big. On going nearer to it, they discovered it was moving, and what they were thinking was buoys on the nets were not buoys at all, and they came to a decision that it must have been a tremendous beast. And some of the lumps were going down, and others coming up, and that led them to understand it was a monster, and an unusual monster, and they tried to get past it – which they managed – and they rowed to the shore as quickly as they could. Well, they were so terrified that they could not even look behind them, but the last look they gave, there were only two bits of the monster showing and then it went down, and never after or before were any of the fishermen working in that place or in that direction seeing it again. And after hearing so much about monsters nowadays it is very probable it was a sea monster – of what nature nobody can say.

Another sea monster

Now the crew of the fishing-boat *Fly-by-Night* was herring fishing and that evening I am going to talk about particularly they were about thirty-three and a quarter miles off Barra Head, on the famous ground called Stanton Bank. There was not a breath of wind and they began to shoot the nets with the assistance of the oars. There was no motor of any description, with the result they had to shoot the nets with their oars. When the nets were shot and everything was still and calm they were on top almost of Stanton Bank and they heard a fearful moaning from a sou'westerly direction and they all looked at each other with astonishment, wondering what was this moaning. Then shortly afterwards they saw at the stern of the boat a tremendous beast breaking the water, and this beast, which was horrible to look at with the size of it, put its paws on the very deck of the stern of the fishing-boat. They all looked at each other and did not know what to say and one of them cried out, 'Get a bucket of fire from the stove.' and they took the bucket of fire and they gave it to the beast right in the head and in the eyes and the moment the beast felt the heat he went down to the very bottom and fortunately he did not make another attempt to put in an appearance.

Had not these men acted so quickly and thrown the bucket of fire on, more than probably the whole boat would have been sunk.

Now they set about hauling the gear. They started to haul their nets – they had about forty or fifty out – and when it was finished there was nothing else for them to do. There was no wind – they had to lie there and still the beast had every opportunity to come to the surface again and attempt to board them, but she did not. And when the wind broke out, they put all the sails they had up, three sails – the fore-sail, the mizzen and the jibsail – and they made for Castlebay as fast as the wind would drive them.

That is the story of the monster that was seen thirty-three and a quarter miles off Barra.

[*For a description of various sea monsters which have been seen at different times, see Willy Ley*, The Lungfish, the Dodo and the Unicorn, *Chapter 7, 'The great Unknown of the Seas.'*]

Fairies, second sight and ghost stories

How time was lost in the fairy knoll

Once upon a time, in the village of Brevig, there lived a man who very often went round the shores to look for any treasures from the sea. In the days I am talking about, shipwrecks were more numerous than the present day.

There is a place round the Ru'Mór called Port an Dùine: this used to be a famous creek for catching whatever might have been blown ashore. Now this day it happened that it was a sou'easterly wind, and a good wind for anything going that would land in Port an Dùine. He stood above the Port, and seeing nothing except a human jawbone with beautiful white teeth. The jawbone drawing his attention very much, he went down to the sea, picked it up and examined it, and said to himself that it was the finest set of teeth that he had ever seen – and at that stage he threw it away, and walked up.

And while he was walking up, a few yards distant on the grass he heard music, pipe music of the finest order; and going closer to it he found a stone there, and when he turned the stone he saw that there was a stairway going down, and he walked down. And there was the piper, dressed in kilts and playing the pipes – an old grey-headed man beautifully dressed in the green, with silver-buckled shoes. As soon as the stranger came in, he was cordially invited to sit up at the fire, and this the man did. And then the music continued.

Food was prepared for him which he did enjoy very much and after what appeared to him to be a short time with the fairy he was told that he could not stay any longer and that he had to go away. And the man walked up the stair on which he came down and was told that when he got up to the top he must turn the stone and put it in the same position as he found it.

He then faced home round the Point, and to the home he left behind him what he thought was a few hours ago. Lo and behold, when he arrived in the house there was nothing in it but bracken, rushes and nettles, and no sign of human habitation at all. That is the very wonderful part of the story – happening in so short a time. Looking about him, he was very much put about and he did not know what to say, but he came to a decision that he must have been inside a fairy knoll. Looking about him he could see no houses except a house a considerable distance from him, and he made a bee-line for the house.

Inside that house was a cobbler repairing boots, in accordance with the custom of that day. The cobbler appeared to be a very old man, and, 'Come in,' he says, 'you are a stranger.'

'Yes, I am a stranger.' he says, and he began to tell him when he left and how he went round the Ru' Mór, the shore, and to Port an Dùine. And then the cobbler halted and began to ransack his mind and first he said that he never heard of the man; but latterly he said, 'Yes, now I remember. I remember seeing my great-grandfather, and he heard from his father than an old man went round the shore at the Ru' Mór and he never came back. And so you must be the man that I heard of from my great grandfather now.' And they left it at that decision, and the man who was in the fairy knoll sat on the end of the bench and he found himself getting very feeble.

Now at that stage he was getting weaker and weaker, and he asked the cobbler to send a message if he could to Eoligarry, to the church that was built in the days of St Columba, and from there came a priest. He gave the man that was in the fairy knoll last rites and as soon as that was done – very peculiar – he crumpled down, a lump of earth.

John the postman and the fairies

I am now going to tell you a fairy story – I am sure one of the last stories about fairies that were seen on Barra. There was a postman in Bolnabodach called John and he was appointed postman after the ferryman stopped crossing the Sound of Barra with letters and the steamer service began to carry the mails. Now his mother was dangerously ill and he was asked to go through the night to fetch a priest, but he was afraid to do it. But, however, as soon as daylight set in, he left his home and went across the hills to Craigston, where the priest was. And it was then getting very nearly sunrise. And when he climbed up to the very top of the hill the sun had risen. He was looking about him and looking towards Heaval, a place called the Glaic Ghlas (the Green Valley) when he saw a lot of little men and women running about in and out of a rock there. First he thought that they were people milking cows and on second thoughts he considered that he was too early in the morning for milking cows. Then they disappeared altogether at one instant and then again they were more numerous than ever; and so that continued for some time until he came to the conclusion, 'Well, they must be fairies,' he says, 'because people are not milking cows at this time of the morning at all.'

He hurriedly went on and made a bee-line for the priest's house at Craigston, and the priest put his head out of the window and asked him where he was going and he told him. 'Well, better go over,' he says, 'and tell the servant to get the pony ready for me, and I won't be very many minutes.' So John told the servant to get the pony ready and he went before the priest as he was sure that the priest would overtake him with the pony.

Now some way on the road he was taken up by the priest and they began to talk and John told the priest what he had seen. And first of all the priest said, 'Well,' he says, 'they might have been people milking cows.'

'And that is what I thought myself,' said John, 'at first, until I saw every one of them disappear and then, all of a sudden, they came out again. But now I am perfectly confident that they were not anybody but fairies.'

And so the priest said, 'Well,' he says, 'it is many years since I heard of any fairies being seen on Barra, and I am satisfied that that was them.' So that is the very last fairies that were seen on the Island of Barra.

[*They have been seen since. For a description of the fairy world, see Kirk's* Secret Commonwealth; *also Wentz,* The Fairy Faith in Celtic Countries, *which includes stories from Barra.*]

The fairy wedding on Hellisay

Once upon a time there was born on the Island of Hellisay a man called Angus. He was one evening standing in the yard among the haystacks and he saw a brilliant light across the water shining, and immediately a whole lot of people began to run in and out of a knoll – so well-dressed that he came to a decision that it must be a wedding; and the music was so wonderful, he never heard anything else to come near it. Sometimes he would see a pair coming out after dancing, talking and more or less appearing to Angus to be sweethearts. Angus was enjoying the scene very much.

Now it was customary at a Highland wedding, when it was getting late, for a woman to be told off to take the bride and a man to take the bridegroom. And when these had gone out, two others would immediately take their places and the dancing was carried on – and this was a position of great pride, to be asked to take their places. So now it was getting near bed-time and the bride was attended by the bridesmaid and the groom was attended by a best man, and then it came to the point where they were sent to bed and that was done. And to complete the custom of the day, there was a big bottle of whisky put below the pillow, and a big lump of cheese on the table – and as far as I could see they were well equipped!

Well, latterly it was very nearly daylight before they dismissed and Angus thought that the whole ceremony took only a matter of minutes. And the music so beautiful, and the dancing so excellent that he was in fairyland without a doubt.

Angus died last year [1951] at the ripe age of 102, and that is his description of a fairy wedding – and the only description ever I heard from any man or woman of seeing such a thing on the Island of Barra.

[*Hellisay is an island off the north-east of Barra. It is no longer inhabited. I collected the place-names of Hellisay in* 1937 *from Murchadh an Eilein, who was born there. He told me of a certain flagstone, that unless a dish of milk was left there the people were liable to find that the fairies had driven their cattle on to an inaccessible promontory.*]

The man taken from Canna by the sluagh *or fairy host*

Once upon a time there was a man on Mingulay called Neil, and he was a great rock climber, and one night he was climbing the biggest cliff in Mingulay and he came against trouble – he came to a point that he could not pass and he then got afraid that if he tried it he might come down the precipices and down into the water. Now he was talking to himself, wondering what to do, and thinking that he was getting hungry, and the thought went through his head, 'Well, be it who it likes, even the Devil, I will volunteer to be his if he takes me out of this difficulty.' And immediately the thought had gone through his head he found himself on the top of the cliff, and the voice told him, 'Now,' it says, 'any time I want you, you will be ready to go with me, for I saved your life to-night, and bear in mind, that means you are mine for ever.'

At the moment Neil was very pleased to close the bargain at that, but afterwards came the trouble. The calls became so numerous and he found himself a slave to the one who took him out of the difficulty. In the end, to avoid his calls he went to America, and even there he was followed by the same routine that he had in Barra.

Now I quote here one instance. Neil was in Canna and he had a call to go to Barra – and up he went – and across he went, and he landed in the Island of Flodday, which is at the back of Fuiay. It was in March, and the woman of the house, the crofter's wife, was cooling gruel in two wooden bowls – pouring from one to the other – and immediately Neil came inside and he asked was there anything to eat, she said, 'Hold on a minute, Neil, until I cool this, and I will give you a bowl of it.' And she gave him barley bread and butter and the bowl of gruel, and Neil started to eat the bread and butter, and drink out of the bowl, and he did not get the job finished when he was called away again.

Away he went and he was landed at Bachd, between Borve and Tangusdale on Barra. Well, he went in and there was a shoemaker there. He knew Neil, too, and he asked Neil where he came from, and was he hungry. And he says, 'No, I'm not hungry – I took a bite of bread and butter and gruel on Flodday.' And he was not long there when a call came again that took him to Mingulay and Neil had calls of this nature for many years before he went to Canada – and as already described, after going to Canada he was followed by the same master.

This story proves that there were aeroplanes of a kind in existence before this present day.

[Falbh air an t-sluagh – *being carried away by the fairy hosts. There is a very good story on this theme told by Eachann Mac Dhubhghaill in* An Ròsarnach, *Vol IV,* p. 41.]

MacAskill and the second sight

In South Uist second sight was more common than in any other part of the Outer Islands. During the 1914 war there was a sailor fellow from South Uist came home on leave and he visited his neighbour. Exchanging news, it was late at night before the sailor made for home, and on the way home he told MacAskill to halt – that there was funeral coming. Now this they did, and MacAskill was standing beside him during the period it took the funeral to pass them.

'Now you can come to the middle of the road – the funeral has passed,' he says. 'Before I come back, such and such a man will be dead. By the appearance of the mourners I am guessing that.'

I gather that after anyone has a vision in that line he generally feels very sick, and he told MacAskill, 'I think I had better sit down because I do not feel very well at all.' They were a considerable period exchanging news, and he said to MacAskill, 'Well, I had better come home with you and stay the night, because I don't feel at all well.' They turned and walked all the way home and he stayed the night along with MacAskill.

Then next morning they began to talk over the matter. 'I have not had a vision of that line for a long time – that is why,' he said, 'I was so very sick after it. And I will be going away in a few days, and strange as it may appear to you, MacAskill, I never see anything of the sort when I leave the soil of South Uist behind me.' He came on his next leave and went to see MacAskill, and they began to talk it over. MacAskill says, 'How is it,' he says, 'that your prediction was true about the man who died?'

'Oh,' he replied, 'my prediction is always true. I guessed it was going to be the man in accordance with the mourners I saw following the funeral.'

And this is only one of the very many stories connected with second sight.

[*See Martin Martin's book on the Western Isles for others.*]

William and the second sight

When second sight was prevalent in South Uist, there was a man called William, noted for seeing sights. One day he was coming across the Barra Sound, ferrying two men who were going to Castlebay and open a shop for the herring fishing summer season. There were in the boat, William, a lady and the two men. Coming across the Sound, William was steering the boat and he saw very clearly and distinctly the tangles from the bottom of the sea round the necks of the two men – a strange and very peculiar sight to see. And when William saw the sight he was very much shocked and he said nothing. In fact he was speechless for a considerable time afterwards.

Now they were coming into Bruernish, and William decided that he would come out of the boat, himself and the lady, and that they would walk to Castlebay. He knew that the two men's lives were to be lost in the near future, but to safeguard himself and the lady, he decided that they would get out of the boat and walk from Bruernish to Castlebay.

Now the boat set off for Castlebay, and they had rounded past the Currachan, and when they were south of the Currachan a considerable distance there came a sharp shower[1] from the west. And people in Rubha Lios saw the boat before the shower but not after it. And so it was noted that the boat sank during that shower. But they thought that the boat might have turned into Castlebay past the Rubha Mór.

William and the lady arrived in Castlebay, but the boat didn't. They were waiting, and waiting, and didn't know what to do. The news went round that the boat didn't come into Castlebay, and next morning very early there was a boat going out to their great lines. And lo and behold, they saw a boat turned upside down, just right on the track they were sailing, and with a man dead on her keel. And they took the man aboard, and took the boat in tow and turned back to Brevig. The man was stripped and taken care of and identified. So William and the lady got the news in Castlebay and they came down, and to their horror found the man that was with them the day before, lying a dead corpse.

That was the sight predicted by William. Now an arrangement was made to take the remains of the man back to South Uist, and all the merchandise in the boat was lost.

I personally knew the lady and William myself. I crossed to Polacharra one late evening and the weather was very doubtful and

[1] *Fras* – squall and shower of rain combined.

124

I met William at the inn. And I was wondering, after hearing this story, was he seeing anything about my neck?

Second sight in Uist

All over Uist second sight was prevalent – not so much nowadays. There was a man in Iochdar who was famous in the whole of Uist for seeing second sight. One night this man was invited to a wedding. On his arrival the wedding party was dining at the table, and strange as it may seem the man saw the bride in her shrouds over her wedding dress. So much did the vision upset him that he collapsed inside the house. He was taken out into the fresh air and after he recovered, he said, 'I am not going back any more.' But he told a party shortly after that in private that he was sure that that lady would not live to the end of the year. And that was so. The girl was dead before the end of the year, as he said. She died in giving birth to her first child.

Shortly after that, he came out of his house one day, and he said: 'Well,' he says, 'I am today seeing a most wonderful sight.' And his neighbour was with him, and he said, 'What is that sight today, Donald?' 'Oh,' Donald said, 'it was a very queer sight – I don't think anybody else in the whole world has seen the sight I am seeing today. I am seeing a weaving loom very high up in the sky, going at a very great speed.'

'Well,' said his neighbour, 'that is impossible, and that is a sight that nobody could believe – a weaving loom that would be going at a big speed high up in the sky.'

'Well, be that as it is,' he says, 'no doubt it is a strange sight – but it is there.'

And later on in years, when the first aeroplane flew over Uist, my friend said, 'Well, that's the loom I saw.'

This man saw many second sights, in fact, too numerous to mention.

The frith *or divination*

In the summer of 1910 I happened to be at Lochboisdale, as a fish salesman. One day I went round exploring the country, on a bicycle, which was not seen very often in those days in Uist. It was a beautiful summer day. I was coming up the Lochboisdale road and I then turned into Garrynamonie, and to my great surprise I heard a voice at the back of a huge boulder surrounded by iris, saying the Hail Mary. And so much so that I stopped in my great astonishment to listen to the voice – and then she popped up, and it was a very old lady with a shawl over her shoulders and another round her head, saying her prayers. And she says to me, 'Are you the Inspector of Poor?'

I says, 'No, my dear lady, but I know the Inspector of Poor all the same.'

'And who are you?' she says.

'Well, I am a Barraman,' says I, 'that came over to the Lochboisdale pier to sell herring.'

'Ah yes,' she says, 'I heard about you selling the herrings and I heard you were very good at it.' Now she turns round and she says, 'And do you know the Inspector of Poor? Ah well,' she says, 'he is a bad man. He never sent me thatch for my house – and you can come in and see it for yourself, that it is in a poor condition.'

Well, I went in to see the lady and her house, and I agreed with her that her house was really in a bad condition. Now I promised her to see the Inspector of Poor and I didn't wait until I saw him – I sent him a strong letter and swore at him upside down! He was a man called Roderick –, and I knew him well. Two or three steamers after that – well, not the first but the next one – I saw bundles of thatch coming off the boat and to satisfy my curiosity I went over and I saw that the bundles were addressed to Roderick –, District Clerk, Gerinish, South Uist.

* * *

And now the next time I went to see the old lady, the house was neat, well-thatched and quite rainproof, and on arriving I went to see the old lady and we had a talk and she thanked me thousands and thousands of times in succession and in fact prayed for me for my kindness in helping her much-needed request. 'Well, next year,' says I, 'when I come back, if it so happens that I do come back, you will be having an increase of your pension.' And so it was – her pension was increased. Next year, when I called on her, she apologised for not believing me when I told her that she would be getting an increase – I think she almost thought that it was me that gave it to her!

Now we began to speak about things in general and she mentioned to me about a *frìth*. I told her I was curious how and what was to be done before you would complete a *frìth,* and she said that a *frìth* was made at daybreak and you had to take note and see what were the beasts that were in your vision – a cow, stirk, sheep – and the backbone of the *frìth* was considered on the animals on the scene at the time. Well, immediately it struck me – had that anything to do with witchcraft? 'I would not do such a thing to you at all – to teach you witchcraft,' said the old lady. And she told me not to be afraid – that she told it to Canon MacDougall, and he made it very clear on the point that there was nothing connected with witchcraft about it. That is the story as I was told, but still there were rhymes that you had to say, which I refused to accept.

When the old lady got the increase of the pension the post office was the old inn at Polacharra, and she used to take a dram to celebrate the old age pension, and I remember one day arriving there and she had just left. Well, perhaps it was just as well!

[*See the* Scots Magazine *of October* 1955, *p.* 64, *for an account of Catriona nighean Eachainn in Glendale, South Uist, who used to make these divinations. The late Fr Allan McDonald wrote down the signs she went by. These were later published, from Fr Allan's notes, by Miss Goodrich Freer, in* Folklore, *Vol XIII, p.* 47.]

The manadh *or forewarning*

In Polacharra, in South Uist, the hotel keeper told me a story, and this is it. 'Any time,' he says, 'that there is a funeral to be in the district, on the socket on which I put the beer cask I can see it sometimes jumping and hear it thumping pretty often. It is customary to have drinks at funerals in Uist, and every time there is a funeral I can hear this noise a few days before. And as soon as the funeral party comes to collect the beer and clear away, I don't hear the *manadh* any more, until there is another funeral. In fact,' he says, 'I heard several times the cart that was going to carry the jar away coming to the house.'

My own experience in person was this. I was sitting with a young lassie and we heard, both of us, distinctly heard a shot. She remarked about it first and said, 'Before the house of MacLean out here was built I heard my mother saying that she often heard sounds like a gunshot' – and the sound we heard was very like a gunshot.

Now it was a long time after that, and I didn't hear anything about it. But unfortunately the girl in question took seriously ill. The doctor said it was galloping consumption that she had. I went to her funeral. On arriving at the house I heard that same shot again, and looking round me I saw they were just nailing down the boards of her coffin. That is one clear point that I heard a *manadh*. I did not collect this from anybody.

[Such experiences used to be frequent.]

The mermaid

In olden days it was a great belief among fishermen that if any would see the *maighdean mhara* [mermaid] it was a serious forecast that they were going to encounter a hurricane and maybe loss of life.

One day there was a boat leaving the port of Brevig on the east side of Barra manned by six men, very capable fishermen, and when they were about two miles off the shore one of them saw a *maighdean mhara* popping out of the water and, the worst luck, between them and the shore. One of them saw it first and he was very dumb, and then another saw it and he let the cat out of the bag and said:

'We had better turn back,' he says, 'I have seen the *maighdean mhara.*'

'So did I,' says Roderick, and they did turn, to make back for Brevig.

Now they were only a couple of miles off the shore and there came a hurricane off the north-west which compelled the men to stow the sails and the mast, and row to the shore, as the boat could not carry a stitch of canvas. And they rowed with all their might and with all their strength.

They were making the weather better as they were proceeding to the shore, and latterly they pulled up and got in touch with the rocks at the Rubha Mór at Brevig. Well, there was a decision that they would come out and leave the boat there, but one of them suggested that seeing they had got a hold of the rocks it would be a disaster to lose the boat and deprive them of their fishing. So they stuck to the boat and put a rope ashore and they remained there until such time as the gust abated, and when the gale was over they immediately rowed back to the port which was not far away from them and did not go to sea again that day.

But when they did go the next day, they met with beautiful weather and plenty of fish, and the good lady *maighdean mhara* never put in an appearance at all.

The belief has faded. Not in my own memory did anyone see the *maighdean mhara,* though I knew a man from the Island of Eriskay who did find a dead *maighdean mhara* on the beach and he described it very fully, and she was not long dead at all. He was remarking how beautiful her hair was, and how very much like a human woman she was except that she was fish from the waist down.

Crodh mara – *sea cattle*

Once upon a time there lived on the Island of Pabbay a crofter, and when he took over this croft he also took over the stock. And the stock consisted of one cow, and for another thing he was well equipped with a *poit dhubh* [illicit still]. He had a famous one. He was getting on beautifully – the old cow had a calf every year, and from the story it did not seem to me that the man had a family at all. He was getting on very well in Pabbay, selling the stock and all.

One night, he and the wife were talking at the fireside.

'Well Mary,' he says, 'I think we had better sell the old cow this year.'

'Och man,' says Mary, 'no, I don't think we will sell it at all. There is plenty of grazing on the island.'

But he insisted on selling the cow.

'Well,' she says, 'Donald, we were very very lucky that the old cow was there before us when we came to the island, and now from that cow we got our stock.'

Well, Donald was keen to sell, and, 'If we don't sell it,' he says, 'we will kill her and salt her and have her for the winter.'

Ah well, the old wife was so fond of the cow she would not hear of that either, but she would rather kill her than see her leave the island altogether.

Unfortunately the day came when Donald was about to kill the cow, and they were sitting at the fire again a few nights afterwards and they fully decided to kill it. Next day they would go down.

Late through the night after they went to bed, they heard a lot of bellowing out in the barns and getting up they discovered that they had not got one single beast – they had all broken loose and left the island. Next day they searched and they searched the island and could not get a sight of any of the stock. Donald came to the mainland of Barra and told the story of his disaster. Then he went round the oldest people in the island to find out what was the matter, and if they could give him enlightenment why he lost the stock. And he discovered that the old cow, very mysteriously, came from a sea-bull who came ashore and had a connection with a cow from whom that cow was descended. And when the old cow heard the conversation between the master and the mistress (for it was customary in those days for the barn to be attached to the dwelling and in some cases under the same roof) she gave one loud bellow, and all the cows that were tied in the barn broke their tethers and went back into the sea, and the island was left derelict of cattle.

Then Donald had to go to the mainland [of Barra] and buy a cow or two himself. And he could well afford it – he had the best *poit dhubh* on the island, and he had a tremendous stock of whisky hidden in a cave in the north-west corner of Pabbay. It was customary in those days for the exciseman to go round, and as Donald was knocking about one day putting another cask in, they caught him and the whole lot that was in the cave that Donald had had there for years maturing. The exciseman showed no mercy but hit the casks with the sledge hammer and commandeered the lucky 'black pot.' And the remains of that pot are on the Island of Pabbay still – when I was there I saw it myself.

*   *   *

Now the island is derelict, nobody living on it. Fifty years ago there came a hurricane from the south–east and the only boat on the island was out on the fishing bank. People on the hills of Mingulay saw the boat swamped and nobody was ever living on the island since.

The water-horse

If you are staying at Northbay, the quickest way to get to the Loch an Eich Uisge is to follow the road by the plantation and go right on till you come to the bridge at Loch an Dùin. Then follow the stream going right up to the hill and that stream will take you to Loch an Eich Uisge. So anyone who will read this story can go on his own direct to the loch with these instructions.

One fine summer day there was a bonny lassie herding her cows at the side of the Loch an Eich Uisge and lo and behold, she saw a most beautiful, charming fellow sleeping. And she went over, and so attractive was his head and the beauty of his hair, 'I had better go and comb it,' she said.

Now the colour of the hair was interesting her very, very much, and when she further went into it she found the reeds that were in the loch, and that surprised her a lot. And she began to fear that he might be the water-horse in the shape of a good-looking man. Now she began to get very much afraid. The man's head was in her lap and immediately an idea struck her to slip off her skirt and leave the skirt below the head and clear out. And that she did, and her tartan plaid was not far away from her and she flung the tartan plaid over her and she ran away and she ran for home.

Well, then, the man woke and turned into a horse and to his great disappointment the lady was gone – and it was fully intended that she would be his victim – and getting into a furious rage, with his hooves he smashed a lot of stones and it is the quarrying he made in his terrible mood that is there to be seen to this day. So after that, never was the water-horse seen on the loch again.

Uaimh an òir, *the cave of gold*

There was always a traditional story in Barra that one could walk from Cliat on the west side of the island to Port an Dùine on the east side of the island. So a party set out, and there were two pipers and three dogs, to explore. And one of the pipers started to play the pipes at Cliat and – very peculiar – those that were following were hearing the pipes and keeping the course until they came to Port an Dùine, and the pipes then faded away and none of the party were ever discovered. How they met their fate no one knows, but the dogs that came out, they came out bare, without a hair from their nose to their tail; and what was the cause of that no one could form an opinion. But it is believed they met some queer folk below the island who killed the two pipers. And so the dogs could not tell what happened, and then they were showing no signs of recovery, the people decided to shoot them. Therefore there is no trace of what happened to the party.

[*This story is told of other places in the Highlands and Islands also. See Frances Tolmie,* Gaelic Folksongs from the Isle of Skye, p. 157.]

The story of the giant's fist at Bagh Hartabhagh

Once upon a time there lived three fishermen on the Island of Eriskay, and they started fishing lobsters – it was early in August. They were not fishing at Eriskay but at Bagh Hartabhagh. Seeing that they were a considerable distance from home, they built a shieling there before they went. Then they started to fish: they got on very well – plenty lobsters and plenty flounders. They were using flounders as bait for the lobsters, for the lobster is very fond of the flounder.

They were getting on very well, and about the 15th August they were going to get new potatoes and this night they decided to rise early in the morning and get lugs for bait and bait the line, so that they would have done the whole job in the one day – that is, got the line down and got the lobster traps hauled and baited. Everything worked out to plan and they baited the line and set it and went to the creels and hauled, and there were a considerable number of lobsters. They went to their lines as they were on their way home and there were a smashing lot of flounders.

There was a neighbour who had a lot of potatoes planted near the shieling, and he gave them permission to go to his lazy beds any time they wanted potatoes and fill their bag and use them. So the next day they thanked him very much, lifted the potatoes and took them home. They made a very early start again and this day they were finished early, and when they came home they boiled the flounders and the potatoes, and the furniture was very scanty in the shieling, so they spread the oilskin on the floor and capsized the flounders and potatoes out on to it. Now they were innocently cracking away, talking about the lobsters – how plentiful they were – and the flounders – how beautiful they were. And there they were, and at the same time going into the tuck that was on the peculiar table on the floor, and in the centre of the enjoyment the floor broke and a huge hand, an enormous hand, came through. And he was beginning to open his claws – the giant – and the three poor men stared at one another, eyes sticking out of their heads with fear, and they cleared out of the shieling and they did not give themselves time to close the door even, but the last fellow to look behind them saw the claws opening in and out. The boat was quite handy, and with all speed they ran down to the boat and made for home. They arrived at Eriskay, very much to the astonishment of the people at home, at arriving at such an unusual hour. Everyone told the story at their own home what happened, and the tremendous giant's hand that came through the floor of the shieling, and so on.

Next day they said to themselves, 'Well, we will go back to the shieling, and if there is nothing wrong we will try it again. When they arrived at the shieling the door was exactly as they left it. They had a consultation and came to a decision that it was just a ghost, and if it came again they would not trouble any more – they would take the creels back to Eriskay. They put on the flounders again and the potatoes and they enjoyed them this night thoroughly and without any interruption. And they continued the whole season right to the end and they did not see the giant fist any more.

[*Bagh Hartabhagh in South Uist is a notorious place for ghosts.*]

Story of the ghost and the plank

Once upon a time there was a man in Eriskay, and at the time I am talking about he was very old, and himself and his son and his grand-son went out fishing to the Oitir. After putting down the lines they went ashore on the Island of Hellisay.

There was a house empty on the island, recently evacuated by a shepherd who was evicted from the island. The old man was keen to get a plank, and peeping through the window he saw one which was used, by the shepherd who had lived there, for a bench in the house. So he decided to open the door, go in and take the plank out. Now he was going to take the plank down to the boat and a ghost met him and tried to take the plank from him – so between the old man and the ghost a struggle started.

The father and the litde boy were in the boat, and so the father told the little boy: 'Calum,' he says, 'go up as quick as you can and see if your grandfather is coming.' So the boy jumped out of the boat and getting near the house he saw the grandfather and the ghost fighting about the plank. As soon as he came within range, the old man bawled out: 'Calum! Calum! Go! Go to the boat and tell your father to come up and take with him the helm of the rudder – and tell him that I am fighting with the Devil.'

And Calum ran as quick as ever he could and he bawled to his father, 'Come! Come! My grandfather is fighting with the Devil!' And the father replied, 'Well,' he says, 'it is high time the Devil came and took your grandfather away – he should have taken him long ago!' So he took the helm with him, the purpose for which he was to use the helm being as a weapon to split the Devil's head. To show you that there was a little of the barbarian in the old fellow, he was going to put up a proper fight.

Now as soon as Donald appeared – Donald was the name of the son – the ghost disappeared. And the old man seemed to be very tired, and he sat down on a rock. 'And why,' he says, 'did you not come at once, before the Devil went away? – for I wanted to give him one with the helm that would knock him useless and senseless and cripple him never to come back again!'

Well, it appears that the ghost had some suspicion that he would have the worst of it when he saw the helm coming. They took the plank down with them to the boat and as they were very nearly at the boat there was an Eriskay boat coming ashore and they talked to one another. And one of the men said, 'Well,' he says, 'this is very peculiar. Last night such and such a woman died in Eriskay and we were all the morning and up to now, round the shore looking for a plank to make a coffin for her and unfortunately we didn't get any – and now that is a plank,' he says, 'that would do beautiful for a coffin. Will you give it to us?'

And Donald says, 'Certainly, yes. I will give you the plank for that purpose.' So the plank changed hands and the men that were in the boat moved away at once and went back to Eriskay with the plank, and they started to make a coffin, and the old lady was put into it and buried, and never again did John see the ghost.

And the mysterious part of the story is this: it is very queer how those Eriskay men failed to get any wood on the Island of Eriskay or round the Island of Hellisay that would do to make a coffin, until they came across the old man that was stealing the plank from the deserted house.

Mary and George

Once upon a time there lived on the Island of Eriskay, Mary, a bonny lassie, aged about nineteen. She fell in love with a sailor who had been away for many years from the island. He was keeping her company for well over a year, and unfortunately it happened that Mary discovered herself to be in a position in the near future of having a baby. And she became very, very unhappy. And it made matters worse when she discovered that Neil left the island without even calling to say goodbye. Mary became very broken-hearted and sick, and went into a decline – so much so that her parents, who did not know what was the matter, decided on sending Mary to some of her friends at Hellisay, to see if she would brush up and get better.

Now this happened to be in May, at the time of the year when each crofter on the island had to keep a look-out to keep the cattle and sheep on the hill, in case they would destroy the crops growing – potatoes, oats, barley and such like. It happened now that the house to which Mary went, it was their turn to do the shepherding the next day, and Mary finding out the position volunteered to get up early, at sunrise, and look at the crops and see whether any sheep or cattle were doing any harm. The people in the house were very much against her doing so, but on no account would Mary obey, and then the mother of the house told her, when she was going out to take with her a bucket, and on returning that she would take home a bucket of water from the well, and she knew where the well was because she was there already the day she came.

Mary went to the well, filled the bucket with water, put it on the track where she would get it when she came back. She made the rounds faithfully and dutifully, and saw that there were no sheep and no cattle among the corn, the oats or the potatoes. When she was on the high, very commanding hillock – which I know very well – lo and behold, all of a sudden there a beautiful lady appeared to her, and when Mary saw the lady, which she knew perfectly well was a stranger on the island, she felt very much afraid. The lady, seeing Mary's thoughts, cautioned her not to be afraid – that she would do her no harm. The lady was a good-looking specimen of a lassie, with a green over-cape over her shoulders, her head beautifully combed and brown hair. Then she broke the news to her that she would have a baby within a period of nine months and that Neil would never marry her. Mary was very much downhearted when she heard that.

At this time there was a full-rigged ship lying out in the Minch, just ahead of her, and Mary was thinking, 'I wish I was aboard that ship, because Neil might be there.' And the lady replied to the thoughts: 'Do not put your thoughts so far away. Neil is not in that ship – and supposing he was, Neil will never marry you.

'Now,' she says, 'Mary, I am going to tell you something of your future career, and anything I forbid you to tell you are not to tell it to any human being on the face of this earth. What I give you permission to tell you are at liberty to do so, any time you so desire. You will be living with your mother, very comfortable, and you will be living on the croft at the harbour of Eriskay, and your two brothers who are abroad sailing, making plenty of money, they shall never forget you – in fact, one of them will come home later on in life.' When the lady was parting with Mary, she again vowed her not to tell anything which she told her not to tell. At the same time she said, 'Goodbye, Mary, one of your race will see me yet.'

After Mary settled down a bit on Hellisay she returned to the Island of Eriskay, and on Eriskay the baby was born. It was a little girl, and her name was Jean. When she grew up, Jean was a good-looking lassie herself. Now Mary used to tell some of the story about the lady she had seen on Hellisay in her younger days, but people were more or less not believing her. And it was not until the next person, George, saw the lady, that Mary began to tell more about it.

After many years passed there was a man George, cousin to Mary, one day cutting rushes on the Island of Hellisay, along with John MacLeod and Dougall MacNeil. They were cutting rushes to thatch their houses, in the month of October. And George, who was a good man with a scythe, was cutting the rushes, while the others were taking it down in bundles to the shore. George was a school-mate of mine, and this is his account of how he saw the lady.

'I was on Hellisay,' he says. 'I was cutting rushes with a scythe, with the fellows I have already mentioned, and,' he says, 'lo and behold, a good-looking lady stood beside me and I got such a terrible shock,' he says, 'at the appearance of the lady that the bonnet fell off my head. And the lady cautioned me, 'Don't be afraid.' Well, Coddy, she began to speak to me and she said to me that I would soon be getting married and that I would have a nice family. And,' he says, 'she told me also that I would never get much of their benefit. Well, Coddy,' he says, 'I thought that they were going to turn out bad.' (But that happened not to be the case. I can vouch with safety that they all turned out very well. Now George was following the life of a sailor, in which capacity he was a very capable man. One night in the North Sea it blew a hurricane and one of the boats broke adrift from the davits, with the result that George was instantaneously killed.)

When George was parting with the lady she said, 'Goodbye, George, one of your race will see me yet. And here's a message,' she says, 'you will take to Mary in Eriskay.' And she gave the message to George, and George faithfully gave the message to Mary. I was with George at his brother's wedding in Eriskay. And at the time we were at the wedding George got an escort right out to the harbour where Mary was staying, to deliver the message. And the wonderful part of the story is that Mary knew that George was coming. So when he arrived at her house, she told the boys who were in her house having a *céilidh* to remove and go outside. Then she took George into her confidence and George told her the message that was sent to her from her old friend the lady, whom she had seen many years before that.

Up to the time of writing, nobody else had the pleasure of seeing the lady.

How Donald met the ghost of Alexander MacDonald, the famous bard

There lived a man in Bruernish named Donald. For some reason or other he had to go to Lochmaddy to give evidence of a drowning accident in Castlebay. Steamers were not so plentiful in those days, and they went over the Sound of Barra and walked from Polacharra to Lochmaddy. Now after the court, they left a message that the same boat would take them home and they were going to meet them on such and such a date in Polacharra.

In Lochmaddy Donald bought a knife – one of the knives that were very common in those days, called the 'cock knife' – there was a picture of a cock on the blade of the knife. Well, Donald went into a house in Carnan, Iochdar, South Uist, as he was married to the first cousin of the people who were staying in the house. First of all they were keen that Donald would stay the whole night, but he said that already he was behind the others and he would have to go in case the boat would go across from Barra to Polacharra and he was not there. And as I knew the man well I can say he was never known to be behind.

Late in the night he started the journey of twenty-one miles on foot. He was at a place called Gerinish when, lo and behold, a very sturdy looking man popped up beside him, and immediately Donald saw him he spoke to him. There was no reply. And so the second move Donald took – he took out the knife and opened it, and said to himself, 'Well, if you make any attempts to speak to me, you will find it will go in to the bird' (about half the length of the knife). And so when he saw the knife the ghost disappeared without any more hesitation, and when Donald saw that, he said to himself, 'Well, there is.no more use for the knife and I'll put it in my pocket.'

When the story went round it was found that it was very common to see the ghost round about there, because he was factor to Mac-Donald of Clanranald. And the very mysterious part of it, he was never seen again, and the story is still alive in South Uist that Mac Maighstir Alasdair was never seen after the Barraman threatened to stab him with the knife.

[*Alexander MacDonald lived from about* 1700 *to* 1770. *He was Baillie of Canna in* 1751, *the year he published his Gaelic songs, which are violently Jacobite in sentiment. Donald is not the only person who has seen his ghost in South Uist, nor the last. His brother, Lachlann MacDonald, had a tack of Dremsdale in the eighteenth century, and it is said his ghost has been seen, too.*]

Witchcraft

The witches who went fishing with a sieve

In Loch Slapin in the Isle of Skye there lived in the village of Slapin a fisherman and a tailor, and they were great neighbours. Their wives were famous witches. One day the fisherman was out all day and he came home in the evening very, very tired, and so he lay down on the bench and fell asleep. And during his slumber who came in but the tailor's wife, and they had a strong talk, and their talk was where would they go fishing to-night. At this period Donald woke up, and slowly he was paying attention to the ladies who were making ready to go to fish, and he was very much amused and very inquisitive to find out what they were going to take with them, and lo and behold, the fisherman's wife asked the tailor's wife, 'What will we take with us?' And the tailor's wife replied, 'We shall take the sieve with us.'

Well, now they were talking away and at last they had everything in readiness to go, and at this moment Donald, lying on the bench, 'woke up' and he asked where were they going. And they said to him, 'You had better lie down and sleep till we come back.' 'Ach,' says Donald, 'I had better go with you.' What interested Donald was when he heard they were going to fish with a sieve, and he insisted that he would get away with them. They point-blank refused him, and they were standing on the floor, the three of them, all ready to go down to the boat. And Donald thought it was a name they had for his new boat, but when they arrived on the beach before they went out they made Donald finally promise that whatever would happen while they were out fishing, he had not to say one single word and especially he had not to use the word of the name of God. And Donald said, 'Anything you tell me,' he says, 'I will agree to it.' Immediately when they heard that they took out the sieve. 'And you will also tell us when we have plenty of herring ashore,' they said, and Donald said, 'Yes, that I will do.'

139

And then immediately they put to sea from the water they became two rats and they moved away gently from the shore and went in a little distance and here came a shoal of herring. A voice from the sieve rolled out, 'Have you got plenty of fish?' And Donald said, 'No, not yet.' Then they moved out to sea and there came more this time of herring ashore. And they rolled out, 'Is that enough, Donald?' Donald said, 'Not yet.'

At this stage they moved out further, and then a huge shoal came and the beaches were covered. And a voice from the sea rolled out, 'Is that enough, Donald?'

And Donald replied, 'Yes, thank God!' – and down went the witches and they were never seen again.

The little witch of Sleat

On the point of Sleat a daughter and her father were one day working on a croft in the early days of summer, and, lo and behold, they saw a ship on the water up past the south end of Eigg. Immediately after, they looked again, and the ship was going to the bottom, and before they stopped looking at it it disappeared out of sight. And he says to the little girl – she was only young – 'I wonder,' he says, 'what was the matter with that ship?'

'I did that,' she said.

'What?', said the father.

'I did that,' she said.

'And who taught you to do such a horrible thing as that – to founder a ship and a crew?' said he.

'Oh, my mother,' she said.

Do you know what he did? He got a spade and killed her instantly when he heard that her mother did teach her it, and he killed the mother and the daughter, and there was nothing said about it. Skye was terrible for witches in those days.

[*Neighbouring districts are often reproached with witchcraft in oral tradition.*]

The Barraman who was bewitched by the woman to whom he gave grazing for her cow

One time there was a fisherman living in Tangusdale, near the loch there, Donald was his name, and his crew were one of the best crews in the island of Barra. During the spring, say March, April, May, June, they were out fishing, and the last fortnight in June and the first fortnight of July they used to go to Glasgow with their fish, fish oil and general cargo. And as there were no steamboats in those days running between Glasgow and the Isles, they did that service with a boat that was built on the Island of Barra.

Donald had a neighbour by the name of Mary, and Mary had a cow, and she had no croft, and before Donald left she asked him could he give her grazing for the cow during his absence in Glasgow, and as soon as ever he came back, if he considered it the thing to do, she would remove the cow off his croft. And he agreed.

One fine summer day they started from Castlebay at sunrise and put out four oars – there was not a breath of wind and they were rowing and singing to themselves most of the day until they got a fine breeze in the evening, and then they put up the two sails and she was going very nicely.

They made for the Crinan Canal, and after arriving at Crinan they proceeded up the Firth of Clyde and landed on the Broomielaw, Glasgow. It did not take them a long time to dispose of their cargo and they sold the whole lot. Then they started to purchase hemp for making lines, hooks and sailing twine for reeling the hooks and all the other commodities, including tobacco, sugar, tea and a jar of whisky, and some clothes on a small scale. Except shirts, all the clothes they were needing came from the wool which was carded, spun, woven and even tailored and made on the island.

Now they took a day or two for themselves, and they would have been going to the pictures, only I suppose there were no such things to go to at that time! Then they started for Barra again. They sailed down the Clyde and they were very happy, and they reached Crinan homeward bound. They got through the canal and then they sailed up and up until they got to the top of the Sound of Mull, and it was about sunrise and the wind was blowing about east-north-east – quite favourable to cross the Minch to Barra.

Donald was feeling very happy and so he said to another member of the crew – Donald his name was too – 'Donald', he says, 'fetch that jar of whisky.' And he served a dram all round and then he got one himself. He took up the glass and said. 'Here's your very good health, boys,' he said, 'and good luck to us and a good passage, and I hope before the sun kisses the western ocean of the shore of the Island of Barra to-night that we will be ashore on our dear island.'

They all agreed with Donald, and felt in very good order. One of them says, 'Ah, Donald,' he says, 'we would better have another one.' And Donald said, 'When we are past Muldonich we will get another one.'

Then the wind dropped and there was not a breath at all. Shortly, you know, they heard the roll of the waves between them and Barra, and here there was just a gale of wind coming down from the west-north-west, and immediately they turned round and went into the harbour at the north end of Coll – it is called Bàgh Còrnaig. They went in there, and through the night the wind abated. Next day they made a start and at the same time when they reached the same spot here came the gale dead between them and Barra. Now, mind you, this was in August, and they kept on trying this until latterly they decided to take the cargo ashore and put it into one of the barns of the men of Còrnaig. And nobody seemed to know what was the matter.

The time was going on and the harvest came, and they did not get across. The potato lifting started and they did not get across. Latterly it got on to the beginning of winter and they did not get home. Then the season for the *cèilidhs* started and they used to follow the neighbours as they would go to the different houses.

They followed the neighbours one night into a house and the fire was in the middle of the floor, and the men came in and sat down. They began to tell the stories, and one was in the middle of a story when the horse belonging to the crofter came home. And he was well covered with hailstones.

Well, Donald was sitting very close to the old lady of the house and she appeared to be very old indeed, and her hair was turning brown with the colour of the peat-smoke – and she spoke to herself when the horse came in with the shower and she said, 'Oh, you dim horse, isn't the *cailleach* who is in Barra to-night playing havoc when you came home and had to take shelter under the roof.' Still the gale was blowing outside.

Donald whispered to her. 'Mistress,' he says, 'I would like very much to have a talk with you.'

'Well, so would I, Donald,' she said, 'I would like very much to have a talk with you. And Donald, when the others go away,' she says, 'you stay behind and have a talk with me.'

Later he said, 'Well, boys, I am not going with you just now. I know my way home and I will sit here with Mistress MacLean.'

Now the coast was clear and the interview began, and she said to Donald, 'Well, Donald,' she says, 'you are a kind-hearted man, and you gave grazing to the cow of the widow, Mistress So-and-So. And it is her,' she says, 'that is keeping you wind-bound on Coll. And if you take my advice, and follow the instructions I give you, I will see you before sunrise ashore on Barra.'

Donald was delighted with the news and the remedy to get home.

'I am going to make a thread for you,' she said. 'First of all I will get my *cuigeal* (distaff)' – and she got hold of the *cuigeal* and she spun about a fathom of thread and she put one knot on it, and then another, and then she put another on it, three in all. And she told Donald, 'Donald,' she says, 'I am now after completing the thread, and under the shade of this thread and in accordance with the instructions I am going to give you, if you follow them you will be ashore in Barra before the lady gets out of bed. Muster your crew,' she says, 'and tell them to put everything in the boat.'

The crew were still outside waiting for Donald, and he said to them, 'Well, look here, boys,' he said, 'I am after having an interview with this old lady inside and she made up for me a *snaithlean,* and if we follow the instructions she gave us we will be in Barra before Mistress So-and-So gets out of bed. You get ready the boat and I will be down after you.' Then Donald went in to bid goodbye to the lady.

He went aboard and said to the crew, 'Well, boys,' he said, 'these are the instructions I got. She told me, the lady, when it was dead calm, to loose one of the knots, and to put up the two sails and then she says if I was complaining that I would carry more sail I could loose the second knot, but on no account to loose the third one – because, "If you do," she says, "it is doubtful if you will ever smell the shore – there will come a hurricane after that."'

They started from the shore and at the time they were rowing; and they let go one of the knots and they got a nice breeze, and it was a late moon and it was getting late at night, and when they caught sight of Muldonich they said to themselves, 'Well, we will loose the second knot.' And they did, with the result that they found a rattling breeze – so much so that Donald was feeling inclined sometimes to ask them to reef the sail, but as he felt in a hurry to get ashore in Barra before the lady got out of bed, he did not.

They got inside the loch at Castlebay – and in fact they got in between the castle and the shore, and Donald said to himself, 'Well, now,' he says, 'I am going to test what actually was the strength in the witch at Coll who did us famous, and who prohibited me from untying the third knot.' And he did it, and a gust came from the northwest which threw the boat, the cargo and the men on the shore – and if he had loosed that when he was on the sea, very probably Donald would never have been seen again.

He had to walk out of the boat and go home. And when he arrived the lady got up and she came to Donald – she had the brass face to come and tell Donald, 'Well, Donald,' she said, 'I am very glad you have come.'

And he said, 'Get out of my house,' he says, 'you witch, that kept me in Coll since July last.' And he swore at her upside down and told her to clear her cow out of sight, and clear out herself and no sympathy or kindness or comfort would he ever show her. From what I am told, the lady left the village and sold the cow, and had no more to do with Tangusdale or Donald.

Note on the second edition

After having been rejected by several publishers, *Told by the Coddy* was published privately in an edition of two thousand copies in March 1960. It is therefore with great pleasure that I record that within a year the demand for it has justified the printing of a second edition.

The Coddy is the only Hebridean storyteller I have known who told his tales with almost equal fluency in both Gaelic and English. I say 'with almost equal fluency' because, as was natural, his style was superior in his native language; but of course only the bilingual reader will be able to compare the two. I must take the opportunity to make clear again that the stories in this book represent the Coddy's *ipsissima verba* as taken down in shorthand by Miss S. J. Lockett in English, as one or two reviewers seem to have been uncertain whether or not they were translations. But in any case, there is no one living who can take down Gaelic in shorthand, and there would have been no point in the tales being taken down in Gaelic shorthand when a wire recorder was already available for the work.

Practically everything that was taken down from the Coddy is included in this book. I have added the Gaelic texts of two more of the stories, which I have transcribed from my wire recordings I made in 1950, to this edition, so that readers may get a further impression of his Gaelic style. It can be seen from these that his English versions are not verbatim translations of these stories so much as the retelling of them in a different language.

J. L. CAMPBELL
Isle of Canna
30 January 1961

Mac Nìll Bharraidh₃ Mac Na Banntraich, Agus am Boc Sealtainneach

(faicibh taobh 39)

Bha mac Banntraich uaireigin dha'n t-saoghal a' fuireach ann am Miu'alaidh. Cha robh aig a mhàthair ach e fhéin, agus nuair a rachadh Mac Nìll a Mhiu'alaidh, bha e gabhail beachd uabhasach math air a' ghille. Thuirt e fhéhin r'a mhàthair: "'S fheàrr dhut," ars esan, "an gille thoirt dhomh, agus togaidh mi fhìn e 'sa chaisteal, agus bidh e air a bhiadhadh, agus bidh e air a chur air dòigh, agus nì mi duin' eireachdail dheth."

"Dà," ars ise, "Mhic Nìll, cha n-eil agamsa ach e fhéin, do mhic no do nigheanan, agus is cruaidh le m' chridhe gun dealaich mi ris."

"O ge tà," ars esan, Mac Nìll, rithe, "ged a dhealaicheas tu ris, chì mi fhìn air do dheagh-chumail thu, agus bheir thusa dhomhsa 'n gille."

"O, falbh an diu mar a tha thu," ars ise, "ach an ath-latha thig thu, bheir mi dhut e."

Co dhiù an latha bha'n seo, cha robhe fad'air falbh, agus chaidh Mac Nìll a Mhiu'alaidh, agus thug e 'n gille dha'n chaisteal. Cha robh e an uair sin ach òg,'s nuair a dh'fhàs e suas ann am bliadhnaichean, bha rud ann a bha Mac Nìll a' faicinn gu robh fìor-choltas an deagh-ghille, agus gur h-e duine sgoinneil a bha dol a bhith ann.

Thòisich iad an sin mu dheireadh, e fhéin's Mac Nill, ri carachd, agus là dha na laithichean, bheil thu faicinn, cha seasadh Mac Nill turus dha idir. Chaidh e sìos air a h-uile turus aige. Bhuail an seo rud eile ann an ceann Mhic Nill, 's thuirt e ris fhéin:

"Tha," ars esan, "am fear-sa dol a dh'fhàs cho làidir, agus ghabh e ormsa cheana, agus gabhaidh e air iomadach fear eile a bharrachd orm; ach fhad's a bhios mise beò, 's math leam gur h-ann agam fhin a bhios an urram gura mi as treise am Barraidh, agus," ars esan, "bho'n a chuireas mise a thogail creiche a' bhiurlainn 's an criú, cuiridh mi innt' e," ars esan, "agus bàthar a h-uile duin' aca. Agus sin mar a chuireas mi crioch air Mac na Banntraich."

146

'S ann mar sin a bha. Thànaig a' bhiurlainn a dh'ionnsaigh a' chladaich am Bàgh a' Chaisteil, agus an déidh dhasan an gairm, Mac Nìll, leis an dùdaich a bha ri gairm orra (chaidh iad innte.) Agus bha am bàgh geal le stoirm an iar-thuath, agus ged nach ligeadh an t-eagal dhaibh a ràdh, cha robh duine air bòrd a bha deònach falbh.

Agus 's e Mac na Banntraich an gille-tòisich a bh'innte, agus, "seo a nis, ma thà," arsa Mac na Banntraich, "beannachd leat, mo charaid Mhic Nìll" – agus rug e air chùl dùirn e, agus shiab e air 'na broinn mar gun deanadh e air gille beag. Nis, dh'fhalbh iad a mach o'n bhàgh, agus cha robh iad fada nuair a dh'iarr Mac Nìll orra tilleadh. Agus thuirt Mac na Banntraich ris: "Cha till sinn idir; ma tha i fiadhaich agaibhse an dràsd, tha i pailt cho fiadhaich againne, agus," ars esan, "'s e bhios ann, cumaidh sinn oirnn." Agus chum iad orra gus an deach iad ann am fasgadh Sloc na h-Iolaire ann an cúl Maol Dòmhnaich, agus thòisich iad air a h-iomaireadh ann a shin gus an deach an t-sìd' uabhasach a bh'ann a sin seachad. Nuair a chaidh, dh'aontaich iad tilleadh a staigh a Bhàgh a' Chaisteil a rithist, agus chaidh iad air tìr, agus ghabh iad an tàmh, agus thug Mac Nìll biadh is deoch is càirdeas dhaibh, ged a bha e dol 'gam bàthadh an dé reimhe sin.

Nist, bha an ùine dol seachad, agus dh'fhalbh Mac Nìll an sin, agus chuir e mach fiadhachadh dh'ionnsaigh a h-uile duine bha air ùir Alba's nach robh, agus aims na h-eileinean, Sealtainn, 's as gach àite, chaidh a chur mu chuairt gu rachadh a mach fiadhachadh air son fear a rachadh ris a shabaid. Fhuair e fiadhachadh, agus 's ann a Sealtainn a fhuair e fiadhachadh, duin' ainmeil a bh'ann a sin, nach robh riamh a leithid 'na fhearann, air an robh am Boc Sealtainneach. Chuir e mach am fiadhachadh. "Théid mi riut," ars esan.

Nist, thànaig an latha a chuireadh air leath air son na sabaid, agus thachair iad ann an Sealtainn. Bha eagal gu leòr air Mac Nìll a' falbh, agus thuirt e ris fhéin: "'S fhàerr dhomh Iain Mac na Banntraich a thoirt learn, air eagal 's gun dean e dad orm." Agus dh'fhalbh Iain comhla ris. Rànaig iad Sealtainn, agus thuirt Mac Nìll, nuair a chunnaic e tighinn e, "Nist," ars esan ri Iain, "ma chì thu mise a' cur mo làmh air mo chùlaibh dà uair, tuigidh tu gur h-e an t-am agad fhéin a dhol a staigh, agus a dhol eadar mi fhìn agus am Boc Sealtainneach." Ach gun sgeul fhad' a dheanamh dha'n té

ghoirid, no té ghoirid a dheanamh dha'n té fhada, thòisich an t-sabaid, agus an t-sabaid gharbh, agus mu dheireadh chunnaic Mac na
Banntraich Mac Nìll a' cur a mach a làimh', agus ghabh e beachd air
agus chunnaic e sin a rithist air a dheanamh. Ghabh e uige agus sheas
e eadar am Boc agus Mac Nill agus thuirt esan ris, "Seasaidh mis'
thu," agus thòisich an uair sin a' chòmhrag agus an cath a b'fhiach
cath a gh-ràdha ris. Cha robh Mac na Banntraich fada gus na chuir
e 'n gaisgeach air a dhruim fodha, agus chuir e glùn air an uchd aige,
agus dh'iarr am Boc fathamas, ligeil leis éirigh 'na sheasamh, agus gun
deanadh iad rérite.

"Ni mi sin," arsa Iain, "ligidh mi 'nad sheasamh thu gun teagamh
ma dh'aidicheas tu gu robh thu air do bheatadh." "O, tha sin furusda
dhomhsa, tha 'n saoghal a' faicinn gu bheil mi air mo bheatadh." lig e
'na sheasamh e, agus nuair a bha e bruidhinn air éirigh 'na sheasamh,
có leum a nail ach Mac Níll, agus tharraing e sgian dubh as a sleagh,
agus bha e dol g'a cur as a shloc. Thionndaidh Mac na Banntraich ris,
agus thuirt e: "Ma chuireas tu sgian dubh no geal an còir an duine,
feiridh mise ort, agus ann an ionad nam bonn, nach fhaic thu Barraidh ri d' mhaireann." Seo mar a bh'ann. Ghabh Mac Nìll an t-eagal,
agus dh'fhàg e 'n t-ait' agus an rud air fad.

Nuair a dh'éirich am Boc, thug e fhéin agus Mac na Banntraich
làmhan air a chéile, agus thuirt am Boc ris: "Bidh tusa ann an Sealtainn comhla riumsa ri d' mhaireann, agus duine sam bith thig bhuat;
agus gheobhadh tu do ragha phòsaidh ann ma thogras tu, agus bidh
thu[1] sona saidhbhir socair comhla riumsa air an t-saoghal-sa, agus,"
ars esan ri Mac Nill, "faodaidh tusa a dhol dhachaigh, a Mhic Nill,
cha bu choir dhut bhith air tighinn an seo idir o nach b'aithne dhut
sabaid a dheanamh taca ris an duine chòir, ris an fhear a chuir mise
air a' chiad *trip* a chaidh mo chur air talamh riamh," ars esan. Agus
dh'fhan Mac na Banntraich còmhla ris a' Bhoc as a' Sealtainn, agus
dh'ionnsaigh an latha an diu gheobh thu Clann Níll ann a thanaig
o'n duine sin.

(2/1/50)

[1] Sic

Fear Shanndraidh Agus an Guirmein

(faicibh taobh 88)

Uaireigin dha'n t-saoghal, agus cha bu mhath an latha sin, dé thachair ach gun do reacadh saghach air taobh an ear Bharraidh aig na Sgeirean Fiaclach, àite glé ghoirid dha'n Churrachan, agus 's e guirmein an cargo a bh'innte, agus cargo fiachail a bh'ann. Cha robh duin' arm an uair ud eòlach air guirmein no air rudan fiachail dha'n t-seòrsa sin, cha robh móran do dh'úidh ac' unnta ann. Ach co dhiù bha fear ann am Barraidh, fear dha na tanaistearan a Chloinn Nìll, agus bha e fuireach ann an Sanndraidh; agus 's e duine mór taitneach a bha as an duine gun teagamh, agus bha Sanndraidh leis, agus bhiodh e daonnan ag iasgach, agus bhiodh e dol a dh'iasgach dha'n chuan agus naoi snaoidhmeannan air a léinidh – léine a bh'aige; agus 's e bhiodh a' tarraing nan lion, agus bhiodh bian craicinn air a bhialaibh, cha robh oillsginean a' dol as an amm a bh'ann.

Nist, chunnaig e seo—agus 's e duine a bh'ann a dh'fhiachadh ris an rud a dhianamh a dh'fhairslicheadh air a h-uile duin' eile—thuirt e ris fhéin "'S fheàrr dhomh falbh a Ghrianaig 's bheir mi liom ann luchd na sgothadh dha'n ghuirmein, agus ma théid liom, 's math e, 's ma theid 'nam aghaidh, is docha nach bi àrach agam air." Dh'fhalbh e. Thug e leis gillean tapaidh còmhla ris, 's luchdaich iad an sgoth le guirmein aig a' Churrachan, agus dh'fhalbh iad a Grianaig. Fad's a ghoirid gun dug iad air an ràthad, co dhiù rànaig iad, agus chreic iad an guirmein fada, fada os cionn an rud a bha dúil aca fhaighinn air. Agus thuirt fear Shanndraidh ris fhéin an uair sin: "Well" ars esan " 's e seo obair tha dol a phàigheadh, agus 'fhearr dhomh stad o iasgach."

'S e bun a bh'a nn, stad e o iasgach, agus ann an amm mathas na bliadhna, thòisich e air tarraing a' ghuirmein, gus mu dheireadh thug e air falbh do ghuirmein, nach robh am margadh air thuar a bhith cho math tuilleadh; agus sin an rud a bh'ann, stad e dheth a tharraing. Agus's e le òr a bhithte 'ga phàigheadh;'s nuair a thànaig e dhachaigh, agus an trip-sa, chùnntais e – cha n-eil fhios a'm do chúnntais e, ach co dhiù, bha lumlan peic' aige do shòbharain òir. Cha robh duine fiosrachach mu bhancaichean no dad an uair sin, agus cha robh ach na bha 'n sin am peic' òir a ghleidheadh.

Dh' fhàs e 'n sin sean, agus thug Mac Nìll a bha ann an latha sin dha Sgurrabhal, agus nuair a dh'fhalbh e gu bhith 'na Fhear Shanndraidh, chaidh e 'na Fhear Sgurrabhail; ach lean am peice fad na h-ùine ris; agus bha am peice a staigh ann an gaoirnlear mór, agus glas agus iuchair air a' ghaoirnlear; agus mar bu trice a h-uile turns a bh'ann, cha rachadh e a mach as an taigh uair 'sa bith ach an iuchair à bha ann a' seo m'a amhaich.

Nist, rinn e 'n sin brògan; rinn e brògan dha fhéin, agus uachdar mhurain unnta, agus buinn do chraiceann caorach; agus sin agaibh na brògan a bhiodh air, nuair a thigeadh e a mach as an taigh a's t-earrach, agus shuidheadh e as a' ghréin aig ceann an taighe. Latha dha na laithichean, dh'éirich e, agus bha e dol a dheanamh a' cheart-rud, agus gu tubaisteach gu dé rinn e ach dh'fhàg e an iuchair a staigh. Agus an déidh dha a bhith treis 'na shuidhe ris a' ghrein, thuirt e ris fhéin " Hud, hud!" ars esan "nach mi bha mobhsgaideach! Dh'fhag mi an diu an iuchair a staigh, agus 's fheàrr dhomh a dhol dh'a h-iarraidh."

Nist, bha e air pòsadh glé bheag do dh'ùine reimhe sin, agus bha a bhean fada, fada, na b'òige na e fhéin; agus nuair a ranaig e staigh, bha i as a' ghaoirnlear, agus cha n-fhaiceadh e ach pìos dha'n druim aice an sàs as a' pheic òir. Agus ghabh e null far an robh i, agus thuirt e rithe "Tha mi smaointinn " ars esan "gu bheil gu leòr agad" agus rug e air gàirdean oirre, agus chrath e 'n gàirdean aice, agus thuit na bha 'na cròig, 'na làimh co dhiù, thuit e ann am broinn a' pheic' air ais.

Cha robh a bhean uabhasach toilichte dheth air son sin a dheanamh, ach co dhiù, shuidh e a staigh agus thuirt e ris fhéin gum b'fheàrr dha rudeigin a dheanamh dha'n pheic' òir; gun do sheall a bhean dha an diu, nuair a gheobhadh i greim air a' pheicidh air fad, nach biodh i fada cur as dha. 'S e rinn e, chuir e a dh'iarraidh an t-sagairt, agus chuir e a dh'iarraidh a' mhinisteir, agus oilltear a' mhinisteir, agus dh'inns e dhaibh an rud a bha dúil aige dhianamh, gu robh dùil aige a thiomnadh a dheanamh; agus na bha dh'òr as a' pheice, gu robh e dol dh'a fhàgail, agus gur h-ann aig daoine bochda bha e dol dh'a fhàgail. Agus thòisich an sagart agus am ministear air bruidhinn cà 'n cuirt' e; agus dh'aontaich iad air a chur ann am banc a bh'ann an Dun Éideann, 's tha e ann fhathast cuideachd, tha mi creidsinn.

Agus sin an cùmhnantan a chuir Mac Níll sios an Sgurrabhal; thuirt e mar seo: "Tha mise" ars esan "a nist a' fàgail na bheil a' seo a dh'òr as a' pheice, an riadh aige a bhith air a riarachadh aig daoine bochda am Barraidh; agus bidh e aig an t-sagart, agus aig a' mhinistear ann am Barraidh ri faicinn gum bi sin air a riarachadh. Agus seo agad" ars esan "fliad 's a bhios e ann. Tha mise a nist a' fagail a' pheic' òir agaibh, agus dol dh'a chur ann am banca, agus an riadh aige ri bhith air a riarachadh air a h-uile duine as feódhmaile ann am Barraidh; agus an uine tha mi cur reimhe" ars esan "seo agad an úine: fhad 's a bhios bainn' aig boinn duibh, no muir a' bualadh ri lic, bidh riadh mo chuid airgid-sa a choisinn mi gu daor, ag obair air a shon, bidh e air a riarachadh air daoine bochda Bharraidh."

Nist, chon an latha an diu, tha sin mar sin. Riadh airgead Fhir Shanndraidh 'ga riarachadh aig na daoine 's feódhmaile na chéile air an cuideachadh air tìr am Barraidh. Sin agaibh a nist naidheachd fhior, a chuala mise aig fear a mhuinntir Bharraidh, cha n-eil e an diu beò agus chuala e fhéin i aig fear facia na bu shine na e fhéin. Sin agaibh mo naidheachd, ma thà.

(2/1/50)

Na Tri Snaoidhmeannan

(faicibh taobh 141)

Bha siod ann reimhid, uaireigin dha'n t-saoghal ann am Barraidh air a'
Leitheig, àite tha ann an Tangusdal, iasgair, Dòmhnall Dòmhnallach,
duine cho tapaidh, e fhéin 's a chriù, 's a bha air ùir Bharraidh. Bha
daoine tapaidh ann am Barraidh, thogte sgothan ann, agus bhiodh
iad ag iasgach leis na sgothan sin, bha sin mun dànaig bàta-smùide,
no bàta 'sa bith eile, no idir aeropleun, mun danaig iad riamh goirid
dha'n eilein. Bhiodh na Barraich a' togail sgothan am Barraidh, chunna mi fhìn fear a bha ris an obair, agus chunna mi bàta a thogadh
cuideachd, agus i air fàs gu math sean.

Bhiodh iad a' dol fad an earraich ag iasgach, ag iasgach gu trang
as a' chuan, iad fhéin a' sailleadh 's a' tiormachadh 's a' cur air dòigh
an éisg; agus bha sin aca deiseal, agus bha iad a' falbh a rithist leis a
Ghlaschu, leis na bh'aca do dh'iasg, air a' cho'la diag mu dheireadh
dha'n t-samhradh agus ann am fìor-thoiseach an fhoghmhair, nuair a
bhiodh an t-sìde math air son an obair a dhianamh.

Ach co dhiù, bha fear a' fuireach ann an Tangasdal, Dòmhnall mac
Dhòmhnaill, agus bha e fhéin agus an criù a bha còmhla ris 'nan daoine cho sgoinneil 's a bha ann am Barraidh as an amm. Bha iad fad an
earraich ag iasgach, a' deanamh deagh-iasgach cuideachd, agus 'ga thi-
ormachadh, agus ga ghlanadh's ga chur air doigh cho math. An sia latha
dha na laithichean bha iad deiseal, 's cho lumlan a mach air a bial; bha
uighean innte, bha iasg innte, bha ol' innte, agus gu dé nach robh innte a'
falbh. Latha dha na laithichean, mar a thuirt mi cheana, bha iad deiseal.

Ach co dhiù, bhà bann-nàbaidh aig Dòmhnall, cailleach, agus
bha aon mhart aice, agus cha robh fearann idir aice. Agus dh'fhoigh-
neachd i do Dhòmhnall nuair a bha e falbh, "Nist, a Dhdmhnaill' ars
ise "tha thu dol a Ghlaschu; agus gun soirbhich thu leat. Agus nam
biodh tu cho math 's gun doireadh tu dhomh fhin fiar dha'n bhoinn
ghlais gus an digeadh tu, bhithinn uabhasach fada 'nad chomain."

Bha Dòmhnall na dhuine còir aig an robh cridhe mòr fialaidh

farsaing co dhiù; thuirt e ris a' chaillich, "Gheobh thu sin, a Mhàiri, gheobh thu fiar dha'n bhoinn gus an dig mise gun teagamh. Bidh sinn a' falbh a nist" ars esan "an diu fhathast, tha mi creidsinn, agus bidh am fiar sin agaibhse gus an dig mise air ais."

Dh'fhalbh iad co dhiù, a togail a mach a Bàgh a' Chaisteil 'ga h-io-maradh, agus gun deò air an t-saoghal; ach eadar iomaradh is seòladh, bha iad a' cumail rompa fad an rathaid gus an d'rànaig iad an Crianan; agus nuair a fhuair iad ro'n Chrianan 's ro' Chaol an t-Snàimh, bha a' chuid bu chunnartaiche dha'n astar tuilleadh seachad. Rànaig iad a' sin mu chuairt 's ghabh iad a suas Cluaidh; rànaig iad an t-àite far am b'àbhaist dhaibh a bhith creic an éisg, agus rud 'sa bith eile a bheire-adh iad leo, 's thòisich iad air creic, agus ri creic gu math cuideachd. Agus ann am beagan ùine, am beagan laithichean, bha am bàt' aca falamh; agus thoisich iad an uair sin air ceannach. Bha iad a' ceannach min, beagan, agus tì, cha robh tì r'a fhaighinn am Barraidh an uair sin, agus siucair, agus rudan feómail eile dhaibh, bheil sibh a' faicinn, a bh'aca, agus cuideachd, bha iad a' ceannach móran do chainb, agus dubhain, agus a h-uile goireas a bhuineadh do dh'iasgach, bha iad 'ga cheannach an uair sin.

An latha bha iad ullamh, thuirt Dòmhnall gu robh e dol suas dha'n bhaile, agus gun ceannaicheadh e galan uisge bheatha a bhiodh aca air a' rathad. Chuala mi am fear a bha 'g innse na naid-heachd ag ràdh, gun d'fhuair e 'n galan air dà thasdan diag. Ge tà, cha dianadh e sin an diu! Dh'fhalbh iad 's thug iad an aghaidh sìos Cluaidh; fhuair iad dha'n Chrianan, agus fhuair iad reimhe, agus co dhiù, bha iad latha dha na laithichean a nuas Caol Muile, agus a' ghrian dìreach ag éirigh as an uisge. Bha dà sheòl aca rithe 's e 'n uidheam seòlaidh a bha riutha, bha seòl-deiridh rithe agus seòl-toi-sich; agus bha i falbh gu math, cho math 's gu robh 'n gnothach a' còrdadh ri Dòmhnall gu h-anabarrach, agus gun duirt e ri fear dha na gillean, "Iain" ars esan "bheir ugam a nuas am pige sin shuas gu faigheamaid dram."

Chuireadh mu chuairt dram, 's thuirt Dòmhnall "Air ur deagh-shlaint', fhearaibh, agus tha mi an dòchas, mum pòg a' ghrian an cuan an iar Bharraidh an nochd, gum bi sinn air tír." Thug e slàinte mu chuairt dhaibh air fad, agus bha iad gu math toilichte le dùrachd Dhòmhnaill 's a h-uile sian, 's bha i falbh, 's bha i falbh gu h-eire-achdail. Ranaig iad a' sin a mach air ceann Cholla, agus nuair a chaidh iad mu ochd no naoi mhíltean 'san iar air Colla, spàrr iad ann am fiath, gun deò air uachdar an t-saoghail; agus an ath-rud a chunnaig iad, 's e stoirm o'n iarthuath a' tighinn eadar iad agus Barraidh. Agus gaoth 'nan ceann, cha robh i freagairt ris a' seòrsa bhàtaichean a bha 'n seo idir, cha robh iad math gu seòladh le gaoth 'nan ceann idir ach glé mhiadhaineach; agus thuirt Dòmhnall air eagal mum millte na bh'ac' innte, gum b'fheàrr dhaibh tilleadh, agus a dhol a staigh do dh'acarsaid Chòrnaig ann an Colla.

Seo mar a bh'ann, thill iad, 's chaidh iad a staigh dha'n acarsaid, agus chaidh iad suas, 's bha na Collaich gu math ciatach riu, agus 's iadsan a bh'ann a' sin. Chomhairlich iad dhaibh gum b'fheàrr dhaibh fuireach gus am màireach. As a' mhionaid gun d'fhuair iad air dòigh a nist, cha do mhair sin fada, dh'atharraich a' ghaoth agus co dhiù, chaidh am feasgar seachad. Thuirt iad riu fhéin, nach fhiachadh iad dhol tarsainn a' chuain gus am faigheadh iad aimsir toiseachadh latha agus gaoth chothromach.

Là'r-na-mhàrach, dh'fhiach iad. Aig a' cheart àite-sa, spàrr iad as a' cheart-fhiath, agus an ath-rud a chunnaig iad, gaoth mhór an iarthuath eadar iad is Barraidh. Thachair sin an diu mar a rinn e an de, agus b'fheudar dhaibh tilleadh a dh'acarsaid Chòrnaig. Bha iad 'n sin a nist; dh'fhiach iad sin iomadach uair, 's dh'fhàs iad mu dheireadh cho searbh bhith 'ga fiachainn, agus an gnothach a' fairsleachainn orra, gun dug iad aiste an cargo air fad, agus gun do stobh iad ann an taighean beag e, aig na Collaich, agus 's ann a thòisich iad air a' bhuain còmhla riu, 's air a' bhuain a chur air a ceangal, 's a' buain, 's cur a staigh an arbhair.

Chuir iad a staigh an t-arbhar, agus bha 'm foghmhair a' teannadh gu chrioch, 's thòisich iad a' sin air a' bhuntàt' a thogail. Bha iad cómhla riu a' togail a' bhuntàta; cha robh guth air faighinn a Bharraidh. Oidhche dha na h-oidhcheannan, bheil thu faicinn, 's am buntàta air thuar a bhith aca a staigh, thuirt iad gun tòisicheadh iad air a dhol air chéilidh mar a bha na Collaich fhéin a' deanamh, agus 's e sin an rud a bh'ann, chaidh iad ann. Chaidh iad a thaigh. Agus bha cailleach, bha seann-té an t-saoghail mhóir an taobh shuas dha'n teine, agus bha Dòmhnall a' gabhail beachd oirre, agus ag aibhseachadh cho sean 's a bha i. Ged nach robh e 'g radha guth rithe, bha e 'g ràdh 'na inntinne fhèin gum b'uabhasach fhein an coltas sean a bh'oirre.

Ann an teis-mhiadhain "greas ort" thànaig fras o'n iarthuath, fras

uabhasach do chlachan meallain. Dé thachair ach thànaig an
t-each, an t-each dubh a staigh, agus sheas e air rniadhain an ùr-
lair, agus thionndaidh e a chùl ris an àird an iarthuath, agus sin a'
chiad bhruidhinn a fhuair a' chailleach. Bhruidhinn i, agus thuirt i
mar seo: "Ùbh, ùbh, ùbh!" ars ise "eich dhuinn, nach e a' chailleach
Bharrach a tha deanamh ort an nochd! Chuir i dhachaigh chean'
thu."

Agus nuair a chuala Dòmhnall seo, thuirt e ris fhéin "Dia, tha fios-
rachadh agadsa nach eil agamsa, agus cha déid e seach an nochd gus
am bidh fhios agamsa dé bha thu ciallachadh mu'n each."

Agus co dhiù, chum iad air naidheachdan, naidheachd a null's
naidheachd a nall, mu dheireadh gus an robh an t-amm ann falbh;
agus dh'fhan Dòmhnall air deireadh. Agus chaidh a h-uile duine a
mhuinntir an taighe a chadal ach a' chailleach agus Dòmhnall. "Nist"
arsa Dòmhnall rithe "chuala mis' thu ag ràdh rud ann a' siod an nochd, agus chuir e móran do dh'annas orm, nuair a thànaig an t-each
dhachaigh nuair a bha 'n fhras ann an uair ud, dé bha sibh a' ciallachadh."

"Ah, a Dhòmhnaill" ars ise "a dhuine bhochd, tha mise làn truais
riutsa agus ris an fheadhainn tha còmhla riut an ùine a tha sibh ri
port ann an Colla. Tha sin agadsa air tailleabh cho fior-choibhneil 's
a bha thu fhéin, agus sin" ars ise "thug thu fiar dha'n bhoinn ghlais,
dha'n chaillich a tha ri'r taobh ann am Barraidh, agus sin an té tha
'gad chumail ri port ann a' seo fhad 's a tha thu ann."

"Da, bhoireannaich" arsa Dòmhnall "thug mise dhi fiar gun teagamh, ach nach uabhasach an camas a th'agam, agus tha mi gabhail gu
bheil fhios agads' air a' sin."

"O, tha fhios agamsa" ars ise "siod an té a tha 'gad chumail-s' ri port
an Colla. Ach" ars ise "ma ghabhas tusa mo chomhairle-sa an nochd,
bheir mi comhairle ort, 's gheobh thu air tìr a Bharraidh ach cha dean
math dhut dad a dhianamh ach an rud a dh'iarras mis' ort. Agus" ars
ise, 's i cur a làmh air a cùlaibh a null's i toirt a nail cuigeal, "tha mise
nist a' dol a dheanamh snàithlein dhutsa, Dhòmhnaill, agus aig a'
cheart-amm tha mi dol a chur tri snaoidhmeannan air an t-snàith-
lein; agus nuair a thogas tu bho chladach Cholla, 's ma bhios a' ghaoth
gann ort, ma bhios tu talach air a' ghaoith, fuasglaidh tu snàithlein, a'
chiad fhear."

Nist, rinn an criù mar a dh'iarr Dòmhnall orra, agus fhuair iad
a h-uile greim a bh'aca a chàradh gu sàbhailte as a' sgothaidh, agus

chuidich na Collaich iad gu math cuideachd. Agus nuair a bha a h-uile sgath a bha 'n sin deiseal, thog iad a mach o'n chladach, agus chuir iad rithe an seòl agus dh'fhuasgail iad fear dha na snàithleinean. (Cha robh nist aca ri—ach a dhà fhuasgladh; ach an treas fear, cha robh aige ri fhuasgladh idir, thug a' chailleach an aithn' ris-san air.)

An déidh dhaibh togail a mach o'n chladach, chuir iad rithe an dà sheòl, agus dh'fhuasgail iad fear dha na snàithleinean, agus leis a' sin, bha i falbh gu riochdail, faodaidh mi radha cho riochdail 's a bha i riamh. Agus nuair a bha iad a' sìor-theannadh a nail gu math ri Barraidh, agus nuair a bha iad air teannadh a nail, bha Dòmhnall a' dol rud bu mhisniche agus an críu cuideachd. Bha deireadh gealaich ann, a' ghaoth cothromach, cho cothromach 's nach b'urra dha bhith na bu chothromaiche. Agus nuair a bha iad a' tighinn glé ghoirid do Mhaol Dòmhnaich, thuirt Dòmhnall riu "Well a nist, fhearaibh, ma tha sibh a' smaointinn gun doir i leatha barrachd,'s gum bi sinn a dhìth tuilleadh gaoitheadh, fuasglaidh mi snàithlein eile."

Bha feadhainn aca deònach, 's bha feadhainn nach robh, ach co dhiù mu dheireadh, smaointich iad air fad gu fuasgailte an dala snaoidhm, agus ma dh'fhuasgail, cha robh iad buileach uile gu léir toilichte; bha i falbh math gu leòr mar a bha i an toiseach. Thànaig iad seachad air Maol Dòmhnaich's thànaig iad a staigh air bial Bàgh a' Chaisteil, agus ca robh iad a' dol a thighinn gu tir ach air taobh a deas a' Chaisteil air Lamaraig na Mònadh,'s e an t-ainm a th'air. Thuirt Dòmhnall ris fhéin "A nist" ars esan "o'n a tha sinn air tìr tuilleadh, agus nach robh an còrr eagail dhuinn faisg air co dhiù, fuasglaidh mi an treas fear." Bha sin a' bristeadh air riaghailtean na caillich Chollaich. Agus dh'fhuasgail Dòmhnall an snaoidhm; agus ma dh'fhuasgail, thànaig beithir as an iarthuath, agus shrad e air tìr an sgoth 'na monasg, agus fhuair Domhnall air éiginn a bheò aiste.

Agus nuair a fhuair e a mach,'s a chuir e nàdur do dhòigh air na rudan a shàbhail e, thug e aghaidh air an taigh. Agus air a' [Leitheig] chunnaig e chailleach, agus ghabh i a nall 'na choinneamh gu stàtail, agus thuirt i ris "Do bheath' an dùthaich, a Dhòmhnaill, nach mi tha duilich fhad 'sa bha sibh ri port!"

"Bheil thu duilich?" arsa Dòmhnall.

"Tha" ars ise.

"Ma thà" ars esan "cha ruig thu leas a bhith duilich tuilleadh.

Togaidh tu leat bó ghlas an fhir mhóir as m'fhianais agus na faiceam-sa t'aghaidh no t'aodann a' tighinn a dh'iarraidh feòir ormsa as do chùl fosgailte air uachdar an t-saoghail."

Cha dànaig a' chailleach a dh'iarraidh fiar air Dòmhnall tuilleadh,'s ged a thigeadh cha doireadh!

(2/1/50)

Bibliography

BORGSTRÖM, PROFESSOR CARL HJ. *The Dialect of Barra in the Outer Hebrides.* Norsk Tidsskrift for Sprogvidenskap, Vol. VIII.

BUCHANAN, DR DONALD. *Reflections on the Isle of Barra.* London, 1942.

CAMPBELL, J. L., MISS ANNIE JOHNSTON, and JOHN MACLEAN, M. A. *Gaelic Folksongs from the Isle of Barra.* Text, translation and five 12-inch discs, from recordings made on the Isle of Barra. Linguaphone Institute, 1950.

CAMPBELL, J. L. *The MacNeils of Barra and the Irish Franciscans.* Innes Review, Vol. V, p. 33.

CAMPBELL, J. L., and CONSTANCE EASTWICK. *The MacNeils of Barra in the Forty-Five.* Article printed in the Casket of Antigonish, Nova Scotia.

CAMPBELL, J. L. *More Notes on the MacNeils of Barra.* The Scottish Genealogist, January 1960.

MACKENZIE, SIR COMPTON, J. L. CAMPBELL, and PROFESSOR CARL HJ. BORGSTRÖM. *The Book of Barra.* London, 1936.

MACLEAN SINCLAIR, REV. A. *The MacNeils of Barra.* Celtic Review, Vol. III, pp. 216–223.

MACLEAN SINCLAIR, REV. A. *Clann Nèill Bharra.* Mac Talla, Vol. II, No. 45, and Vol. III, No. 20.

MACNEIL, JOHN. *Dunnchadh Ciobair an Ceann Bharraidh.* Gairm, Vol. II, p. 271.

MACNEIL, MICHAEL, and MICHAEL BUCHANAN. *The MacNeils of Barra.* Oban Weekly News, 24th April, 1907.

MACNEIL, R. L. *The Clan MacNeil.* New York, 1923.

MARTIN, MARTIN. A description of the Western Islands of Scotland. Edited by Donald J. Macleod. Stirling, 1934. (Description of Barra and adjacent islands, of Kismul Castle and the *gocman,* etc. pp. 156–164.)

Minutes of Evidence of the Crofters' Commission (Barra witnesses, Vol. I, pp. 643–698; South Uist witnesses, Vol. I, pp. 698–749.) Edinburgh, 1884.

O'LOCHLAINN, COLM. *Deoch Slàinte nan Gillean, Dòrnan òran a Barraidh.* Dublin, 1948.

WILSON, JAMES. *A Voyage round the Coasts of Scotland and the Isles.* Edinburgh, 1842. (Description of Barra, pp. 459–486.)

ADDITIONAL PUBLICATIONS

CAMPBELL, J. L., and COLLINSON, F. *Hebridean Folksongs,* Oxford, 1969.

CAMPBELL, J. L., and HALL, TREVOR H., *Strange Things.* The Enquiry by the Society for Psychical Research into Second Sight in the Scottish Highlands, the story of Ada Goodrich Freer, the Ballechin House ghost hunt, and the stories and folklore collected by Fr Allan McDonald of Eriskay. London, 1968.

CAMPBELL, J. L., *Songs remembered in Exile, Traditional Gaelic Songs from Nova Scotia recorded in Cape Breton and Antigonish in 1937, with an account of the causes of Hebridean Emigration, 1770–1835.*

GIBSON, JOHN S. *Ships of the '45.* London, 1967.

MACNEIL, R. L. *Castle in the Sea.* London, 1964.

MURRAY, W. H. *The Hebrides.* London, 1966.

After the second edition of this book was published, the late Robert Lister MacNeil, who restored Kismul Castle, was granted the undifferenced arms of the MacNeils of Barra by the Lord Lyon in 1963.

A grant of £100 by the McCaig Trust towards the cost of producing the third edition of this book is acknowledged with much gratitude.

MS SOURCES

Justiciary Records: Books of Adjournal, June 3rd, 1678 and July 4th, 1682. John MacLeod of Dunvegan *v* Rorie Mcneill of Bara, James Mcneill his brother and Donald Gair.

Public Records of Scotland: Register of Tailzies, Book 60, Folio 47. Petition by Ewen Cameron McNeil and others for the recording in the Record of Tailzies of the Deed of Entail executed by Roderick McNeil of Barra on the tenth of February 1806.

Public Records of Scotland: Extract of Deed of Settlement by Roderick McNeil Esq., of Barra in favour of Trustees. Registered 7th May 1822.

Scots College, Rome. Letters written by the Rev. John Chisholm, Bornish, South Uist, the Rev. Neil McDonald, Barra, and Hugh MacNeil of Vatersay, to the Rev. Angus MacDonald, Rector of the Scots College at Rome, between 4th February 1827 and 4th March 1831.

Australian Nights: Longing for Summer

HELEN LACEY

MARGARET WAY

KANDY SHEPHERD

MILLS & BOON

All rights reserved including the right of reproduction in whole or in part in any form. This edition is published by arrangement with Harlequin Books S.A.

This is a work of fiction. Names, characters, places, locations and incidents are purely fictional and bear no relationship to any real life individuals, living or dead, or to any actual places, business establishments, locations, events or incidents. Any resemblance is entirely coincidental.

This book is sold subject to the condition that it shall not, by way of trade or otherwise, be lent, resold, hired out or otherwise circulated without the prior consent of the publisher in any form of binding or cover other than that in which it is published and without a similar condition including this condition being imposed on the subsequent purchaser.

® and TM are trademarks owned and used by the trademark owner and/or its licensee. Trademarks marked with ® are registered with the United Kingdom Patent Office and/or the Office for Harmonisation in the Internal Market and in other countries.

First Published in Great Britain 2020
By Mills & Boon, an imprint of HarperCollins*Publishers*
1 London Bridge Street, London, SE1 9GF

AUSTRALIAN NIGHTS: LONGING FOR SUMMER
© 2020 Harlequin Books S.A.

His-and-Hers Family © 2013 Helen Lacey
Wealthy Australian, Secret Son © 2011 Margaret Way
The Summer They Never Forgot © 2014 Kandy Carpenter

ISBN: 978-0-263-29885-7

MIX
Paper from
responsible sources
FSC® C007454

This book is produced from independently certified FSC™ paper to ensure responsible forest management.

For more information visit: www.harpercollins.co.uk/green

Printed and bound in Spain
by CPI, Barcelona